W9-CMM-095

NATURAL VEGETATION
of
South Carolina

*Publication of this book
was made possible by the*
WOFFORD B. AND LOUISE P. CAMP FOUNDATION

NATURAL VEGETATION
of South Carolina

BY JOHN M. BARRY

UNIVERSITY OF SOUTH CAROLINA PRESS

FIRST EDITION
Published in Columbia, S.C., by the
University of South Carolina Press, 1980

Manufactured in the United States of America
Book Design by Larry E. Hirst

Library of Congress Cataloging in Publication Data
Barry, John M.   1944–
   Natural vegetation of South Carolina.
   Bibliography: p.
   Includes index.
   1. Botany—South Carolina—Ecology.
I. Title.
QK185.B28   581.9'757   79– 19678
ISBN 0– 87249– 384– 9
ISBN 0– 87249– 214– 1 pbk.

TO MOLLY

Contents

Preface

Most residents of South Carolina are familiar with our botanical heritage in one way or another, whether it be from picnicking, hunting, fishing, or just plain admiring the scenery. We all know of the spring splendor of the various public and private gardens; however, only a relatively few people have developed an equal appreciation for our botanical heritage. Our natural vegetation has been here longer than any one of us, yet we often take this natural splendor for granted. We have carelessly littered the roadsides and have destroyed vegetation through carelessness and deliberate acts.

The South has long been an area of high cultivation, and with our now rapid growth in population and industrial activity, we should become even more aware of our natural botanical heritage and try to preserve those areas that reflect the past proliferation of our native plant life. The once unique combination of plant migrations, climatic conditions, and noninterference from man and his ever-destructive forces probably never will be duplicated. We must therefore look to the future and try to understand our present condition and protect what remains of the natural vegetation.

The material on the pages that follow is the result of two factors: first, my own curiosity and love for our natural heritage, especially our botanical heritage, and second, my hope that through this work, I may help others to understand and appreciate the great floristic diversity within the State.

I have found it difficult to write a book of this type for many reasons. First of all, it is hard to condense such a diverse topic as our natural vegetation into a manageable form. Second, to walk the middle of the road between the professional botanist (who wants everything to be written in a professional manner and to have every bit of information annotated with extensive bibliographical references) and the non-professional (who, on the other hand, would be quite overwhelmed and often "turned off" by such professional antics) is a feat even more perplexing and difficult than the material itself. Yet, like a true "hard-headed" college professor, I have tried to accomplish this amazing feat, sometimes even to my own dismay and frustration. The manuscript was written basically in a textbook fashion, with references cited at strategic places where specific work has been done. This is not to say that references not cited are less notable, nor have I deliberately tried to ignore any. I have included many more references at the end of the book than are actually cited, because I felt this would be helpful in further investigation by interested persons.

There is an old addage which says "a picture is worth a thousand words." Well, in this work it is worth even more! I can sit down all day and write descriptions about habitats, the dominant plants and their relationships to the environment. But, unless the reader can see an actual example, he is often doing nothing more than reading a species list. Therefore, I have tried to illustrate the principal habitat types by appropriate photographs.

I have also tried to generalize the descriptions of specific habitat types as to the dominant species present. There is one important thing to remember, these are just what they are meant to be, *generalized* descriptions. Very seldom will you find any two sites with the same exact species composition because of varying habitat conditions.

All plants by virture of their genetic makeup have the ability to survive under a certain range of conditions. This fact is easily recognized when we think of the different ecological conditions existing in a desert in New Mexico and those conditions on an island in the South Pacific. Yet, the same principles exist here in South Carolina. Certain plants are able to endure wide ecological variations, while others, because of their different genetic composition, can survive only under a narrow range of conditions. It is these different "genetic tolerances" that cause our vegetation to be reflective of the habitat. Ecological conditions do not often change drastically within a very short distance, and as a result, species composition of habitats may often overlap, reflecting gradual habitat change and variances in individual species genetic makeup. I have tried to support the idea of continuous variation, both in respect to time and space as it reflects local ecological conditions, and still provide enough definitive characteristics for each situation so as to allow recognition as an entity.

I realize that the preceding is quite nebulous, but I wanted to present the material in such a fashion as to reach the greatest diversity of background knowledge. Those who wish to "split" or to "lump" my categories may do so with my blessing. I have noted in the past that botanists tend to prefer to make their own interpretations involving ecological situations, and to this I say "to each his own."

I am deeply indebted to a large number of people; however, I frankly need to single out two as being very significant. Wade T. Batson, Professor of Biology, University of South Carolina, and Earl L. Core, Professor Emeritus of Biology, West Virginia University. Through close association with these two men, I feel that I acquired immeasurable knowledge that is not written in any textbook.

Locally, I feel indebted to numerous people. I would like

to single out Tom Kohlsaat, formerly of the Nature Conservancy and now Naturalist for the Heritage Trust Program in South Carolina; John E. Fairey of Clemson; C. Leland Rodgers of Furman; Richard Porcher of The Citadel; George P. Sawyer of Coker College; Ross Clark of Erskine College; Albert E. Radford of the University of North Carolina; and three former University of South Carolina biology graduate students, John Clonts, Alan Crandall, and Steve Larson. Finally, I would like to thank the South Carolina Department of Wildlife and Marine Resources, my wife Molly, and all of my friends and colleagues too innumerable to mention.

J.M.B.

PART ONE: Introduction

Climate
Physiography

Climate

SEVERAL MAJOR FACTORS combine to give South Carolina a pleasant, mild and humid climate. The State is located at a relatively low latitude (32° to 35° North) and most of it is under 1,000 feet (304.8 m) in elevation. It has a long coastline along which moves the warm Gulf Stream current, and the mountains to the north and west block or delay many cold air masses approaching from those directions. Even the deep cold air masses that cross the mountains rapidly are warmed somewhat as the air is heated by compression before it descends on the southeastern side. This effect can be seen on the maps of minimum temperature in January and to a lesser degree in July, where a fairly large area of relatively higher temperature appears just southeast of the mountains. It is convenient for climatic discussion to divide the State into areas coinciding closely with topographic features.

Elevation, latitude and distance inland from the coast are three factors that work in conjunction with each other to affect South Carolina's climate. Lower temperatures can be expected in the upper piedmont and mountain region, where latitude, elevation and distance inland all have large numerical values; whereas higher temperatures, resulting from smaller values of the three factors, are found along the southern coast. Except for small-scale and local irregularities, there is a gradual decrease in annual average temperature north-westward from 68° F (20° C) at the coast to 58° F (14° C) at the edge of the mountains. Within the Mountain Province, variations in elevation are great over short horizontal distances, and temperature variations are due almost entirely to elevation differences.

All of the record low wintertime temperatures have been set in the Mountain Province or extreme upper piedmont. The lowest temperature on record was -13° F (-25° C) on January 6, 1940, at Long Creek, which has an elevation of nearly 2,000 feet (608 m). The highest summertime temperatures ever observed are not found along the south coast mainly because the ocean waters have very small daily and annual changes in temperature when compared with the land surface. The air over the coastal water is cooler than the air over the land in summer and warmer in winter, and this has a controlling effect on the temperatures of locations on and very near the coast. The July average maximum temperature map reveals that the highest temperatures of 92° F to 93° F (33° C to 34° C) are found in the central part of the State with the coast being 4° F to 5° F (about 2° C) cooler. The July maximum and minimum temperature maps viewed together show a daily range along the coast of about 13° F (7° C), whereas the range is about 21° F (12° C) in the center of the State. Even in January at the time of minimum temperatures, the air over the land a short distance from the coast is 2° F to 3° F (about 1° C) warmer than at the coastal stations. The daily range along the coast in January is about 16° F (-9° C) and about 23° F (-5° C) in the center of the State. The highest temperature on record, 111° F (44° C), occurred three times: at Calhoun Falls on September 8, 1925; at Blackville on September 4, 1925; and at Camden on June 28, 1954.

The growing season for most cultivated crops is limited by the fall and spring freezes. The average length of the freeze-free period varies from about 200 days in the coldest

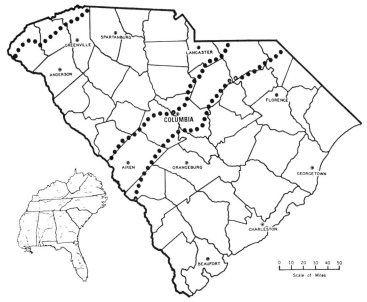

FIGURE 1. South Carolina is divided roughly into four physiographic provinces (from west to east): Blue Ridge Province, Piedmont Province, Sandhill Province, and Coastal Plain Province. The dotted lines are only rough estimates of borders, and local extensions of each province do occur. For example, rolling sandhills (as opposed to the fall line sandhills) occur even as far as Orangeburg where they end at the Citronelle Escarpment.

areas to about 280 days along the south coast. In areas where most of the major crops are grown, it is from 210 to 235 days, or roughly 7 to 8 months. The average date of the last freezing temperature in spring ranges from March 10 in the south to April 1 in the north; the average date of the first fall freeze ranges from late October in the north to November 20 in the south. Freezes have occurred as much as four weeks later than the average date in spring and three weeks earlier than the average date in fall. The minimum temperature is 32° F (0° C) or less on 50 to 70 days in the upper piedmont and 10 days near the coast. Counties in the inner coastal plain and sandhills area have maximum temperatures of 90° F (32° C) or more on 80 summer days, with 30 such days occurring along the coast and 10 to 30 in the mountains.

Clouds and rainfall rarely have more than minor effects on temperature, although maximum temperatures in summer are reduced slightly in areas where afternoon cloudiness and rain are persistent. Another minor cooling effect is drainage of cold air, mostly in winter, into some of the river and stream valleys causing minimum temperatures to be somewhat lower than they would be otherwise. One example of this takes place in a rather deep section of the Broad River valley from Lockhart to a short distance north of Columbia.

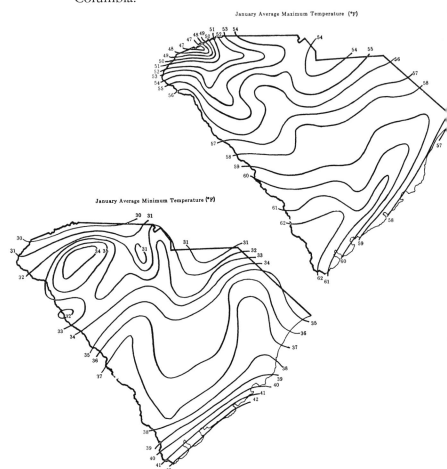

January Average Maximum Temperature (°F)

January Average Minimum Temperature (°F)

July Average Maximum Temperature (°F)

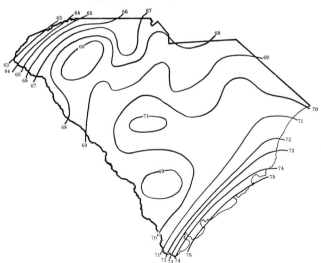

FIGURE 2. Average maximum and minimum temperatures for South Carolina based on U.S. Weather Bureau data from 1935 to 1964. Caution should be used in interpolating these maps and the indicated isolines, particularly in mountainous areas.

Annual rainfall is adequate in all parts of the State. Highest averages of up to 80 inches (203 cm) are found in the upper mountains, whereas the lowest averages of less than 42 inches (106 cm) are found in parts of the inner coastal plain and sandhills. Although there is a decrease in annual precipitation from the mountains to the coast, there are noticeable local and regional variances. The inner coastal plain is decidedly drier than either the outer coastal plain or the sandhills. Such local variances include a dry strip along the southern coast from Edisto Island to Savannah Beach.

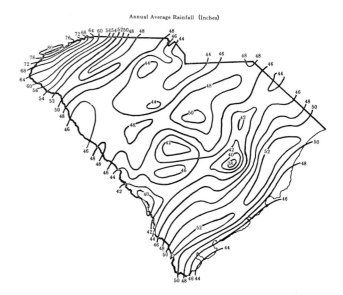

Annual Average Rainfall (Inches)

FIGURE 3. Rainfall is generally adequate over the entire State. There is relatively high rainfall in the Sandhill Province except for a small dry area embedded in it a few miles south of Columbia at the junction of Lexington, Calhoun and Orangeburg counties. Another small dry area is found in the coastal plain around the Santee-Cooper spillway near the junction of Berkeley, Williamsburg, and Clarendon counties. The data were taken from U.S. Weather Bureau information for the years 1935 to 1964, and care should be taken in applying exact locations to isolines.

Statewide seasonal precipitation variances are most noticeable in summer, when definite zones are more easily recognized. In other seasons these zones are less noticeable. Fall is a dry season with amounts dropping to less than 2 inches (5 cm) along the coast by November, although the northeast coast may get additional amounts from tropical storms. Winter has little regional fluctuations, but does show a gradual decrease from 6 to 7 inches (15 to 18 cm) in the mountains to 3.5 inches (9 cm) along the coast. Spring, especially the month of March, has heavy rain in all parts of the State, ranging from 4 inches (10 cm) in the coastal plain to more than 7.5 inches (19 cm) in the mountains. In addition, the pattern of a relatively dry lower piedmont and a relatively wet sandhills begins to appear in April and continues into May. The greatest monthly rainfall on record was 31.13 inches (79 cm) at Kingstree in July 1916; the greatest amount in a 24-hour period was 13.25 inches (34 cm) in the same month and year at nearby Effingham.

In summary, the driest period is in October and November when there is little cyclonic storm activity and "Indian Summer" prevails. Rainfall increases gradually and reaches a peak in March when cyclonic activity and cold front activity are at a maximum. There is a general decrease in rainfall to a dry period from late April through early June. From the latter part of June through early September is a wet period caused by thunderstorm and shower activity that reaches its peak in July, the wettest summer month. The summer maximum stretches a little into the fall along the coast because of occasional tropical storms.

Solid forms of precipitation include snow, sleet, and hail. Hail is not too frequent, but it does occur with spring thunderstorms from March through early May. These thunderstorms usually accompany squall lines or cold fronts. Snow and sleet may occur separately or together or mixed with rain during the winter months of December through February. Snow may occur from one to three times in winter, but accumulations seldom remain very long on the ground except in the mountains. Statewide snows of notable amounts can occur when cyclonic storms move northeastward along or just off the coast. Three intense storms of this type have brought record snowfalls, with a belt of greatest amounts running along the sandhills and the southern edge of the lower piedmont. These storms, in February 1899, 1914, and 1973, blanketed the State with depths ranging from less than an inch to 24.9 inches (63.2 cm) at Rimini in 1973 (Purvis and Rampey, 1975).

Freezing rain also occurs from one to three times per winter in the northern half of the State. This rain, which freezes on contact with the ground and other objects, can cause hazardous driving conditions and breakage of limbs and tree tops and various types of utility wires and poles. One of the most severe cases of ice accumulation from freezing rain was in February 1969 in several north central and northeastern counties. Timber losses were disastrous and power and telephone services were seriously disrupted over a large area.

Prevailing surface winds tend to be either from northeast or southwest because of the presence and orientation of the Appalachian Mountains. Winds of all directions occur throughout the State during the year, but the prevailing directions by seasons are: spring—southwest, summer—south and southwest, autumn—northeast, and in winter—northeast and southwest. Average surface wind speeds for all months range between 6 and 10 m.p.h. (9.7 and 16.1 km/hr).

Severe weather also comes to South Carolina occasionally in the form of violent thunderstorms, tornadoes, and hurricanes. Although thunderstorms are common in the summer months, the especially violent ones generally accompany the squall lines and active cold fronts of spring. Generally, they bring high winds, hail, considerable lightning, and occasionally a tornado. Sixty percent of the tornadoes occur from March through June, April being the peak month with 25 percent. A smaller number arrive in

August and September, accounting for 21 percent of the total.

Tropical storms or hurricanes affect the State about one year out of two, with the majority being tropical storms that do little damage and often bring rains at a time when they are needed. Most of the hurricanes affect only the outer coastal plain; and if these storms do come inland, they lose intensity quite rapidly. The most devastating hurricane, as far as loss of life is concerned, was the one that struck south of Savannah on August 27, 1893. The fast moving storm piled up vast amounts of water to the east of its center, and the south coast and sea islands were badly inundated. Winds of 120 m.p.h. (193.2 km/hr) were measured at Charleston and were probably higher between Charleston and Beaufort. More than 1,000 persons were drowned and damage estimates exceeded $10 million. The greatest amount of property damage ever done, $27 million, was at Myrtle Beach on October 15, 1954. No lives were lost even though the highest windspeed was measured at 100 m.p.h. (161 km/hr).

A hurricane that crossed the State moved inland between Charleston and Savannah on September 29, 1959, and continued on a northerly track. Coastal winds were estimated at 140 m.p.h. (225.4 km/hr), damage estimated at $20 million, and seven lives were lost. Considerable flooding usually accompanies hurricanes that come very far inland, and high tides occur along the coast to the north and east of the center.

Although there have been many earth tremors in South Carolina over the years, there has been only one major earthquake recorded in the State in the last 100 years. After having light tremors on August 27, 28, and 30, a major shock hit the Charleston area on August 31, 1886, just before 10:00 o'clock in the evening. The area of greatest intensity of the shock was about 10 miles west-northwest of the center of the city of Charleston. More than 100 buildings, including frame houses, were demolished. Ninety percent of all brick structures suffered damage, and 14,000 chimneys were destroyed. Eighty-three persons were killed and many injured. The major shock was felt over an area of 2,800,000 square miles, from Canada to the Gulf of Mexico, and from Bermuda to Iowa, Missouri, and Arkansas. Windows were broken even in Milwaukee (Landers, 1970).

Physiography

THE BLUE RIDGE PROVINCE is the belt of mountains west of the Piedmont Province. These mountains are remnants of former highlands that antedated the lower peneplains on either side (Fenneman, 1938). The mountains of this area may be classed as "subdued," that word being used technically to designate a stage in the cycle when height and steepness are so far lost that only a mantle of decayed rock remains over underlying bedrock. Talus slopes and bare cliffs (although they do exist within the State) are rare. Summits are commonly rounded, and although rounded domes are quite evident, true mountain peaks are few. Compared to the Rocky Mountains, the Southern Appalachians are not high. They are considerably more humid, and as a consequence, the weathering process has reduced the peaks to rounded domes.

The rocks that make up the province include Precambrian granite and gneiss. In addition to this basement complex, the rocks in the southern part of the Blue Ridge Province include a thick series of late Precambrian sedimentary rocks consisting of poorly sorted siltstone, sandstone, and conglomerate that grades upward into the Cambrian formations (Hunt, 1967).

In South Carolina the Blue Ridge Province comprises only the northwestern parts of Pickens, Greenville, and Oconee counties. Local mineral resources include small, scattered deposits of gold, silver, lead, mica, feldspar, asbestos, granite, marble, and clay (Johnson, 1964).

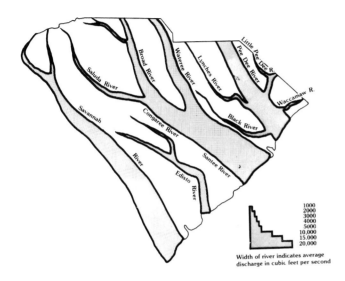

FIGURE 4. Average discharge of the principal river systems of South Carolina (*redrawn from Colquhoun, 1969*).

According to Odom and Fullagar (1973) one of the major problems of Southern Appalachian geology concerns the relationship between the Blue Ridge and Piedmont Provinces. It is not surprising that much work has been done relative to this boundary zone called the Brevard Zone. In fact, Odom and Fullagar say that there seems to have been almost as many interpretations of this fault as there have been investigators. In examining publications by Griffin (1967), Hatcher (1970, 1972, 1975), Rodgers (1972), Haselton (1974), Butler (1971), Birkhead (1973), Odom and Fullagar (1973), and Glover and Sinha (1973), it seems that a possible theory of piedmont formation is that the outer piedmont, including the Carolina slate belt (and possibly the inner piedmont also), was formed from volcanic activity at sea; then because of continental movement, it smashed into the continental margin causing deformations, buckling, and further changes in the metamorphic rock.

The Piedmont Province itself comprises the area from the foothills of the mountains to the "fall line" near the center of the State, an area which represents the change from igneous and metamorphic rocks of the piedmont to unconsolidated sediments of the coastal plain (Johnson, 1964). This dissected peneplain surface slopes from elevations of about 1000 feet (304 m) in the northwest to about 300 feet (91.2 m) along the southeast boundary. Topographically, it is characterized by rolling, highly dissected, piedmont terrain surmounted by monadnocks that give way northward to northeast-southwest oriented ridges and valleys similar in aspect to the physiographic province west of the Blue Ridge Mountains (Haselton, 1974).

Most of the rocks of the piedmont are gneiss and shist, with some marble and quartzite, and according to Haselton, amphibolite units occasionally are responsible for the low saddles or notches in the mountain areas such as the pronounced break in the ridge crest between Table Mountain and Pinnacle Mountain.

Some less intensively metamorphosed rocks, including considerable slate, occur along the eastern part of the province from southern Virginia to Georgia. This area, called the Carolina slate belt, makes up about 20 percent of the province. The rocks in the slate belt are somewhat less resistant to erosion than are the neighboring formations, and they form slightly lower ground with wider valleys. Consequently the slate belt was favored for reservoir sites on the Saluda River above Columbia, and on the Savannah River above Augusta, Georgia (Hunt, 1967; Johnson, 1972).

Mineral resources consist of granite, vermiculite, kyanite, barite, gold, silver, copper, sericite, manganese, asbestos, topaz, pyrophyllite, and shale (Johnson, 1964).

The Coastal Plain Province extends from the fall line to the sea and is composed of nearly flat-lying, unconsolidated sands and clays and soft limestone ranging in age from Cretaceous to recent (i.e., 90 million to less than 10,000 years old). Elevations range from nearly 400 feet (121.6 m) in the central part of the State to sea level at the coast. The coastal plain is roughly divided at an elevation of about 220 feet (66.9 m) near Orangeburg, Shaw Field, Bishopville, and Bennettsville by the Orangeburg Scarp (Siple, 1975), an ancient shoreline (probably Oligocene or Miocene in age, i.e., about 30 million to 20 million years old). Northwest of this scarp is the hilly, upper coastal plain, much of which is known as the sandhills area. Southeast of the Orangeburg Scarp the relatively flat surface of the lower coastal plain extends as an almost featureless plain to the sea. Mineral resources consist of kaolin, sand, gravel, bentonite, "fullers earth," common clay, limestone, phosphate, "coquina," and peat (Johnson, 1964).

The formation of the Carolina Coastal Plain Province is as interesting as that of the Piedmont and Blue Ridge Provinces. Probably the classic work on this interesting concept was done by Cooke (1936). According to him, the seashore has repeatedly shifted back and forth for considerable distances across the Atlantic coastal plain, and all of the present coastal plain is covered by sediments that were laid down either in the sea or on land not far from the seashore. These wanderings of the seashore were due to two causes: first, tilting or warping of the lands, and second, fluctuations of the sea level. With each lowering of the sea level, the coastal plain has been subjected to erosion; but with each temporary stand of the sea, the waves cut back into the headlands and formed bars across the mouths of bays, and these wave-built features remained to mark the land as coastal terraces. In fact, tidal marches and beaches are parts of the most recent terrace. Colquhoun (1969) also cites the lack of secondary barrier islands and marshes along the lower coast as evidence that the present shoreline is of an erosional nature and largely primary in type.

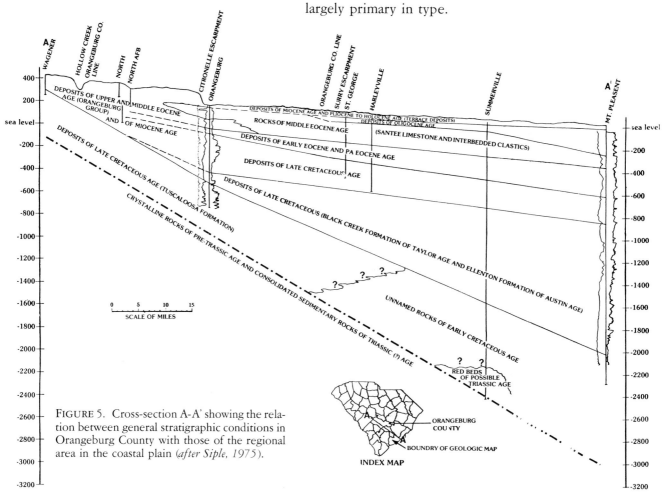

FIGURE 5. Cross-section A-A' showing the relation between general stratigraphic conditions in Orangeburg County with those of the regional area in the coastal plain (*after Siple, 1975*).

PART TWO: Blue Ridge Province

Physiographic features
>   Blue Ridge Escarpment
>   Gorges
>   Interior Plateaus and Valleys
>   Monadnocks
>   Foothills
>   Rock Bluffs
>   Waterfalls
>   Flat Rock Communities

Natural Vegetation System
>   Riverbanks and Alder Zones
>   Floodplain Forests
>   Mixed Mesophytic—Cove Segregate
>   Mixed Mesophytic—Slope Segregate
>   Ridgetops and Upland Oak Forests
>   Pine Forests
>   Rock Communities

Physiographic Features

THE BLUE RIDGE PROVINCE of South Carolina is limited to the northern sections of Greenville, Pickens, and Oconee counties, and is also included in the drainage area of the southeastern Blue Ridge Escarpment. Within the area of this south-facing embayment, the amount of rainfall is the greatest in the eastern United States, and six major rivers drain from the divide. These rivers, the Chattooga, Whitewater, Thompson, Horsepasture, Toxaway, and Eastatoe, descend from about 3000 feet (912 m) at the shelf of the escarpment to about 900 feet (273.6 m) in the gorge mouths and flow over a series of cascades and spectacular waterfalls in their descent of nearly five miles to the South Carolina piedmont. Some of these spectacular cascades however, have been destroyed by the construction of Lake Jocassee. This same lake has completely inundated the Horsepasture and Toxaway rivers within the State and has partially destroyed the Thompson and Whitewater rivers. Even with partial destruction of these habitats, there are still many beautiful gorges to be found, and many are easily accessible to the public.

For hundreds of years the mountains of the Southern Appalachians have been studied by such noted explorers and botanists as Hernando DeSoto, William Bartram, and André Michaux. It is fairly definite that DeSoto passed through the upper part of South Carolina, but his actual route is some-

what uncertain. One account, found in a book by Peattie (1943), suggests that DeSoto marched from Georgia to the headwaters of the Keowee River and then northeastward across the Saluda, Tyger, and Pacolet rivers to the point where the Broad River emerges from the mountains. This would locate him in areas now known as Oconee, Pickens, Greenville, and Spartanburg counties.

Swanton (1946), on the other hand, suggests that the explorer only came into contact with South Carolina near Xuala, an Indian town in what is now northwestern Oconee County. Even if one accepts the longer stay, it remains doubtful as to the impact that DeSoto's explorers had on the flora of the region; so let us consider that this region's first real botanical observations were made by William Bartram during the spring of 1775. Harper (1958) suggests that Bartram camped at what is now Oconee Station in upper Oconee County, and from there moved westward across the Chattooga.

FIGURE 6. Eastatoe Gap from Old Highway 11 in Pickens County. Eastatoe Creek drops from about 2200 (668 m) feet at the North Carolina line to about 1000 feet (304 m) where it comes out into this narrow open floodplain. The distance covered during this descent is about five miles (8 km) and for the most part, the stream flows due south. A number of tributaries enter the ravine from the east and west, with several flowing through large ravines with sizable waterfalls and spectacular scenery. Although much of the area has at one time been lumbered, there are still many mixed mesophytic rhododendron covered slopes and coves containing large hemlocks (*Tsuga canadensis*) and tulip-poplars (*Liriodendron tulipifera*). Most slopes exhibit an excellent transition from mixed mesophytic-cove segregate to slope segregate forests.

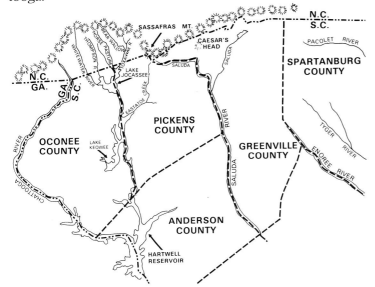

FIGURE 7. Major river systems draining the Blue Ridge escarpment (*adapted from Holt, 1970*).

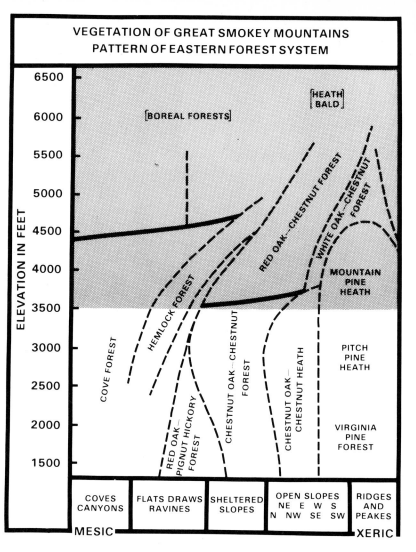

FIGURE 8.   Mosaic chart of the climax vegetation pattern in the South Carolina mountains below 3500 feet (1064 m). The shaded area above 3500 feet is not found in the State (*from Whittaker, 1951*).

The third major explorer of the region was André Michaux in the latter part of 1788. Sargent (1886) states that Michaux traveled in the area of the Keowee River, which is more familiar these days as the junction of the Horsepasture and Toxaway Rivers—or even more familiar, the upper part of Lake Jocassee. If this is true, and it seems to be so, it would place him in the general vicinity of northeastern Oconee County and northwestern Pickens County when he discovered *Shortia galacifolia*.

The richness and diversity of our Mountain Province flora are astounding, yet its evolution is relatively stable. Cain (1943) possibly best sums this up in one sentence where he says: "It is not known whether the Great Smoky Mountains' flora is richer in ancient plant species than other North Temperate regions, because comparable data are not available, but all evidence indicates that the cove hardwood forests of the Southern Appalachians, which have their maximum development in the Smokies, are very similar to the rich, mesophytic, and once circumboreal Arctotertiary forest."

Many slopes of this rich mesophytic forest were dominated originally by an oak-chestnut association. However, sometime around 1904 a blight was introduced into the United States, and it swept over the country from southern New England to Georgia, nearly eliminating the American chestnut (*Castanea dentata*). Remaining dead trees, some still standing, virtually littered the hillsides in many areas. Sprouts still arise from the roots of fallen trees, but these sprouts never attain much size before succumbing to the disease. It might be noted that the decline of the chestnut was gradual enough to let associated species maintain dominance, even to the exclusion of others.

With this in mind, it must be remembered that any classification scheme will be a gross simplification of the

	Mixed Mesophytic Forest: cove segr.	Mixed Mesophytic Forest: slope segr.	Upland Oaks and Ridgetops: Sub-xeric	xeric	Pine Forests: Pine-Oak segr.	Pine segr.
Beech	X	X				
White ash	X	X				
Butternut	X	X				
Sweet Gum	X	X				
Tulip-poplar	D	X	X			
Fraser magnolia	X	X	X			
Basswood	X	X				
Hemlock	D	X				
Red maple	CD	X	D	X		
Dogwood	X	X	X	X		
Red oak	X	X	D	X		
White oak	X	X	X	D or CD		
Pignut hickory		X		X		
Mockernut hickory		X		X		
Black gum		X	X	X		
Chestnut oak		X	D	D or CD	X	X
Chestnut	*		X	X		
Black oak	*	X	X	CD		
Sourwood	*	X	D	X	X	X
Scarlet oak	*		D	CD	D	X
Black locust	*		X	X	X	X
Virginia pine	*	*			X	D
Shortleaf pine	*	*			D	X
Pitch pine	*	*			D	D

FIGURE 9.

General summary of species composition of major vegetation types in the Blue Ridge Province. **D**—Dominant; **CD**—Co-dominant; **X**—Present regularly but usually not a dominant; *****—Never occurs; **No symbol** —May be present but not a typical member of the type.   (Modified after Holt, 1970)

complexity present, and obviously, the vegetation does not occur in neat, well-defined "packets." It does exist, however, as a continuum in relation to several edaphic conditions. Probably the best attempt to describe this continuum in relation to the varying topography was done by Whittaker (1951, 1956). In interpreting his results, along with personal observations, we find that, in the broadest sense, the vegetation of the South Carolina mountains consists of a mixed mesophytic forest in the coves and at elevations below 200 feet (60.8 m); an oak-chestnut (now largely chestnut oak-oak) and oak-hickory subxeric forest at higher elevations; and a pine-oak and pine forest along dry ridges and slopes. We also find that shrubs in mesic habitats are dominated by deciduous ericaceous and non-ericaceous species, evergreen ericaceous species on oak-dominated slopes and ridges, and deciduous ericaceous species in xeric communities.

The following scheme of classifying the vegetation types largely follows that of most area botanists. It is a scheme that tries to divide the continuum into easily recognizable units, but one that tries to avoid making the complex even more complex. The system reflects vegetation types found in the foothills, on monadocks, on slopes bordering coves, and on elevated ridges. The lowest elevations are discussed first, intermediate next, and the highest elevations, with specific communities such as rock communities last.

Natural Vegetation System

Riverbanks and Alder Zones

Many streams and small rivers running through coves may have a narrow border zone only a few yards wide characterized by a dominance of adlers (*Alnus serrulata*). Soils in such border zones are usually rocky with smaller particles sorted by alluvial forces into sandbars and clay banks. Since most of these rivers are youthful, complete stabilization of this zone is not accomplished (Rodgers, 1965). The Chattooga River along the Georgia border has a very gentle gradient compared to most mountain rivers of the Southern Appalachians. Because of this gentle slope, many areas have well-developed alder zones.

In most alder zones, only three trees are normally present to any degree, and even these do not form a closed canopy. The three species most commonly found are sycamore (*Platanus occidentalis*), sweetgum (*Liquidambar styraciflua*), and persimmon (*Diospyros virginiana*). Associated species include blackgum (*Nyssa sylvatica*), ash (*Fraxinus americana*), white pine (*Pinus strobus*), and several species normally found in coves, such as hemlock (*Tsuga canadensis*), red maple (*Acer rubrum*), musclewood (*Carpinus caroliniana*), and tulip-poplar (*Liriodendron tulipifera*).

Cooper (1963) also found that small flat areas adjacent to the alder zone may occasionally support fragments of woody communities. These communities are dominated by hemlock, tulip-poplar, and birch (at higher elevations).

FIGURE 10. A riverbank alder zone along the Chattooga River. The dominant shrub is alder (*Alnus serrulata*), but there is an influx of rhododendron (*R. maximum*), mountain laurel (*Kalmia latifolia*), and small birches (*Betula lenta*). Although the water current is normally gentle, it can be exceptionally strong during heavy rainfall. Soils along the border are rocky, with smaller alluvial particles sorted into periodic sandbars and clay banks. Frequently high currents will completely devastate these accumulations, and their permanence is an exception, not the rule (*courtesy of S.C.W.M.R. Department*).

The most common shrubby species in alder zones include yellow root (*Xanthorhiza simplicissima*), Virginia willow (*Itea virginica*) and sticky azalea (*Rhododendron viscosum*). Other less common shrubs include black willow (*Salix nigra*) and sweet pepperbush (*Clethra acuminata*). Ground cover usually consists of sedges (*Carex* spp.), cinquefoil (*Potentilla canadensis*), several species of violets (*Viola* spp.), and a variety of liverworts and mosses.

Floodplain Forests

Floodplain forests are normally limited to small acreages between the alder zone and gorge slopes. Unfortunately for botanists, most landowners have permitted logging to take place, sometimes even to an alarming degree. In some locations logging has resulted in clearcutting, followed by agriculturization of the land. Those small areas now in forests have existed for only a relatively short period of time; nevertheless, some general trends toward a climax type can be noted.

As to be expected in areas not in a near climax condition, no one or two trees are distinctive, yet trends indicating a mesic to cove hardwood condition do exist.

Tree species may be grouped into two categories, conifers and hardwoods. Principal conifers include shortleaf pine (*Pinus echinata*), scrub pine or Virginia pine (*P. virginiana*), and white pine (*P. strobus*) at low elevations, and more hemlock (*Tsuga canadensis*) and less shortleaf pine at higher elevations. Hardwood species most commonly found are sweetgum (*Liquidambar styraciflua*), tulip-poplar (*Liriodendron tulipifera*), sourwood (*Oxydendrum arboreum*), and red maple (*Acer rubrum*). It might be interesting to mention that while investigating floodplain forests, C. L. Rodgers (1965) found only an occasional oak or hickory. His observation was not totally unexpected since the majority of species present are wind disseminated, as opposed to the slower dissemination of oaks and hickories.

The understory is normally composed of dogwood (*Cornus florida*), musclewood (*Carpinus caroliniana*), sassafras (*Sassafras albidum*), and birch (*Betula lenta*). Shrubs include species such as spicebush (*Lindera benzoin*), strawberry bush (*Euonymous americanus*) and wild hydrangea (*Hydrangea arborescens*). A number of ericaceous shrubs are also present. These include mountain laurel (*Kalmia latifolia*), leucothoe (*Leucothoe axillaris* var. *editorum*), gooseberry (*Vaccinium stamineum*), and another blueberry (*V. vacillans*).

Vines are principally the same as those found on the adjacent slopes, as are many of the herbaceous species. However, there are a few herbaceous species present that we would not expect to find in coves or on slopes. These include common grapefern (*Botrychium dissectum*) and winged sumac (*Rhus copallina*). Species such as spotted wintergreen (*Chimaphila maculata*), partridge berry (*Mitchella repens*), various violets (*Viola* spp.), and Robin's plantain (*Erigeron pulchellus*) are common nearly everywhere in the region.

Mixed Mesophytic Forests—Cove Segregate

As a result of an abundance of moisture and normally good drainage, the flora of coves usually is more luxuriant than that of surrounding sites. The canopy is normally closed, but there is a dense undergrowth of shrubs and herbaceous plants. Spring is especially beautiful since the warm sunlight stimulates the growth of many flowering annuals.

In most cases the two most distinctive species present are hemlock (*Tsuga canadensis*) and tulip-poplar (*Liriodendron tulipifera*). Other canopy species are basswood (*Tilia heterophylla*), umbrella tree (*Magnolia fraseri*), cucumber tree (*M. acuminata*), red oak (*Quercus rubra*), and to a lesser

FIGURE 11. A mixed mesophytic forest-cove segregate community. This scene is typical of many coves dominated by hemlock (*Tsuga canadensis*) and tulip-poplar (*Liriodendron tulipifera*). The canopy is usually closed with a dense growth of large rhododendrons (*R. maximum*) and mountain laurel (*Kalmia latifolia*) along the streams. Away from the canopy opening there is a less dense cover of shrubs and a high concentration of herbaceous plants. Understory includes buckeye, maple, dogwood, and silverbell (*Halesia carolina*) (*courtesy of S.C.W.M.R. Department*).

degree, butternut (*Juglans cinera*), and beech (*Fagus grandifolia*). Cherry birch (*Betula lenta*) also may be evident in the canopy, but it is usually found at higher elevations.

Subcanopy and understory species are numerous. Chief ones, in addition to small individuals of the canopy layer, include buckeye (*Aesculus octandra*), red maple (*Acer rubrum*), dogwood (*Cornus florida*), papaw (*Asimina triloba*), holly (*Ilex opaca*), and musclewood (*Carpinus caroliniana*). Also occurring are Hercules' club (*Aralia spinosa*), and silverbell (*Halesia carolina*). Shrubs present to a high degree include strawberry bush (*Euonymous americanus*), mockorange (*Philadelphus inodorus*), copious amounts of leucothoe (*Leucothoe axillaris* var. *editorum*), rhododendron (*Rhododendron maximum*), spicebush (*Lindera benzoin*), and sweet-shrub (*Calycanthus floridus*). In addition, hydrangea (*Hydrangeo arborescens*) and a semi-parasite called buffalo-nut (*Pyrularia pubera*) are found locally abundant.

Numerous vines are evident, with the principal ones being Virginia creeper (*Parthenocissus quinquefolia*), briers (*Rubus* spp.), numerous species of greenbrier (*Smilax* spp.), muscadine (*Vitis rotundifolia*), and climbing hydrangea (*Decumaria barbara*). It goes without saying that poison ivy (*Rhus radicans*) is present; in some areas it even seems to be the only ground cover present.

Herbaceous plants are extremely evident during spring and again in late summer. It would be completely out of the scope of this work to try to document even half of the herbaceous species encountered, therefore keep in mind that those mentioned are only representative and not inclusive. The most noticeable include species of violets (*Viola* spp.), partridge berry (*Mitchella repens*), foam flower (*Tiarella cordifolia*), numerous species of buttercups (*Ranunculus* spp.), stinging nettle (*Urtica dioica*), sedges (*Carex* spp.), and Robin's plantain (*Erigeron pulchellus*). Other species often

FIGURE 12. Mixed mesophytic-cove segregate. Many coves are not dominated by hemlock; rather there is a higher hardwood diversity. Tulip-poplar still has a high frequency, but red maple (*Acer rubrum*), beech (*Fagus grandifolia*), oaks (*Quercus* spp.), and basswood (*Tilia heterophylla*) increase in importance. There is also a higher frequency of umbrella tree (*Magnolia fraseri*), cucumber tree (*M. acuminata*), and butternut (*Juglans cinerea*). The current can be devastating at times to new growths of aquatic plants and bordering shrubs. Those that do survive are usually in sheltered locations, such as the shrubs behind the large boulder on the right.

seen include jack-in-the-pulpit (*Arisaema triphyllum*), bedstraw (*Galium* spp.), false Solomon's seal (*Smilacina racemosa*), and heart leaf (*Hexastylis virginica*). Two other species very evident include doll's eyes (*Actaea pachypoda*) and lion's foot (*Prenanthes* spp.). The former is recognizable in late summer by its brilliant white berries, each with a black spot and reddish peduncle or stem; the latter by its unquestionable wide variance of leaf shapes.

Ferns are also very numerous, with Christmas fern (*Polystichum acrostichoides*), lady fern (*Athyrium asplenioides*), broad beech-fern (*Thelypteris hexagonoptera*), rattlesnake fern (*Botrychium virginianum*), and maiden-hair fern (*Adiantum pedatum*) usually being the most common.

Mixed Mesophytic Forests—Slope Segregate

As one gets away from the immediate cove segregate of the mixed mesophytic forest and goes up the slopes, there is a gradual change to less hemlock and a greater degree of hardwood domination. This hardwood domination is also accompanied by a greater diversity of species. In addition, we find that the understory and shrub layers are less dense. Herbaceous diversity, however, is increased. The richness of the slope flora results from the additive presence of species characteristic of coves and those more common in oak forests of higher, more exposed slopes (Holt, 1970).

The general observation that there is little zonal segregation is reinforced by the data from other workers of the region. However, there is a general change in dominant composition on north and east slopes from that of south and west slopes, and this change is accentuated by an increase in the steepness of the slope. On slopes such as these, exposure and drying conditions tend to have significant effects on the species composition.

Naturally, it is very difficult to characterize such a diverse community by naming a few species. Rather, it is easier, or at least more conceptual, to single out some of the more prevalent dominants and then let each community site speak for itself on its peculiarities.

The canopy for the most part is dominated by a few very tall species with gaps being filled in by others. Some of the taller species include pignut hickory (*Carya glabra*) and mockernut hickory (*C. tomentosa*), tulip-poplar (*Liriodendron tulipifera*), and several oaks. The oaks most prevalent include chestnut oak (*Quercus prinus*), black oak (*Q. velutina*), red oak (*Q. rubra*), and white oak (*Q. alba*). Other canopy species usually include black gum (*Nyssa sylvatica*), white ash (*Fraxinus americana*), umbrella tree (*Magnolia fraseri*), red maple (*Acer rubrum*), and beech (*Fagus grandifolia*). There may be some appearance of scarlet oak (*Q. coccinea*), but it is usually more dominant on ridge tops. Occasionally one may also find large individual specimens of buckeye (*Aesculus octandra*) and basswood (*Tilia heterophylla*). Understory is relatively sparse and is usually characterized by dogwood (*Cornus florida*), sourwood (*Oxydendrum arboreum*), and black locust (*Robinia pseudo-acacia*). Small individuals of the canopy species are also likely to be present along with small scrubby American chestnuts (*Castanea dentata*). The chestnuts might even be better placed in the shrub layer, but because of their historical size, it is rather hard to completely demote them to shrubs.

Shrubs usually are not as common here as in the more mesic areas near rivers and in coves. Also, there is a gradual change to more xeric species from the mesic species of the coves. On mesic sites shrub dominants include great laurel (*Rhododendron maximum*) and Carolina laurel (*R. minus*), leucothoe (*Leucothoe axillaris* var. *editorum*), and mountain laurel (*Kalmia latifolia*). Drier sites are usually dominated by

FIGURE 13. Reedy Cove Falls in Pickens County. Reedy Cove lies between Twisting Pine Mountain and Rich Mountain and drops approximately 600 feet (182.4 m) in elevation. It opens into a common floodplain with Eastatoe Gap where Reedy Cove Creek eventually merges with Eastatoe Creek. Along its descent, the creek passes through a mixed mesophytic-slope segregate forest. Dominants are usually red maple (*Acer rubrum*), oaks (*Quercus alba*, and *Q. rubra*), basswood (*Tilia heterophylla*) and hickories (*Carya glabra* and *C. tomentosa*), with tulip-poplar (*Liriodendron tulipifera*) and hemlock (*Tsuga canadensis*) occurring sporadically.

FIGURE 14. Mixed mesophytic forest-slope segregate from a cliff above Jones Gap in Greenville County. Such areas were dominated by American chestnut (*Castanea dentata*), but since the blight took its toll, chestnut dominance has been replaced by a mixed mesophytic oak-hickory dominance, with chestnut sprouts rarely attaining more than shrub size. Several large dead chestnut trees (with forked branches) are still standing here, but many others litter the ground along the slope.

buffalo nut (*Pyrularia pubera*), huckleberry (*Gaylussacia ursina*), sweet-shrub (*Calycanthus floridus*), and hydrangea (*Hydrangea arborescens*).

Vines usually are not prevalent enough to become a traffic hazard (except for possibly greenbrier), but several do grow to be quite large. Larger ones include muscadine (*Vitis rotundifolia*) and dutchman's pipe (*Aristolochia macrophylla*), along with smaller species such as Virginia creeper (*Parthenocissus quinquefolia*) and several species of greenbriers (usually *Smilax rotundifolia* and *S. glauca*). Poison ivy (*Rhus radicans*) is commonplace also, but never develops to the extent that it does in the coastal plain.

Herbaceous species are most evident during the early spring before deciduous species develop leaves. Principal ones include violets (*Viola* spp.), Robin's plantain (*Erigeron pulchellus*), Solomon's seal (*Polygonatum biflorum*), false Solomon's seal (*Smilacina racemosa*), jack-in-the-pulpit (*Arisaema triphyllum*), green and gold (*Chrysogonum virginianum*), and foam flower (*Tiarella cordifolia*). Other herbaceous species found include briers (*Rubus* spp.), aster (*Aster divaricatus*), snakeroot (*Sanicula canadensis*), buttercup (*Ranunculus recurvatus*), panic grass (*Panicum*, spp.), hawkweed (*Hieracium venosum*), rosin-weed (*Silphium compositum*), and wood nettle (*Laportea canadensis*).

Ferns are not overly abundant in reference to species diversity, but the frequency of certain species is rather high. Those most commonly found include Christmas fern (*Polystichum acrostichoides*), New York fern (*Thelypteris noveboracensis*), broad beech-fern (*T. hexagonoptera*), rattlesnake fern (*Botrychium virginianum*), and maiden-hair fern (*Adiantum pedatum*).

FIGURE 15. A white pine ecotone on a north-east-facing slope along S.C. Highway 11. The white pines (*Pinus strobus*) occupy an ecotonal area from the stream margin to midway on the slope. A few large hemlocks (*Tsuga canadensis*) with a rhododendron (*Rhododendron maximum*) shrub layer are in the left foreground. White pine and hemlocks are replaced by shortleaf pine (*Pinus echinata*) and hardwoods on the drier upper slopes and ridge top.

FIGURE 16. Extensive mats of running-pine (*Lycopodium flabelliforme*) are not common, but smaller colonies do occur on dry slopes and pinelands in the mountains, foothills, and piedmont.

FIGURE 17. A xeric southwest-facing slope, a rocky area dominated by scarlet oak (*Quercus coccinea*), post oak (*Q. stellata*), and shortleaf pine (*Pinus echinata*). Also present are white oak (*Q. alba*), red maple (*Acer rubrum*), and pignut hickory (*Carya glabra*). Understory and shrub layers are principally sourwood (*Oxydendrum arboreum*), persimmon (*Diospyros virginiana*), dogwood (*Cornus florida*), and smooth sumac (*Rhus glabra*). Such xeric areas are very common on ridge tops and other rocky southwest-facing or west-facing slopes of the foothills and mountains.

Ridgetops and Upland Oak Forests

Most of the upper ridge slopes and ridge tops are dominated by various species of oaks. These oak forests are not uniform and several subtypes may be recognized, if you are inclined to do so. However, the majority of sites may be divided into one of two types: very dry sites dominated by scarlet oak (*Quercus coccinea*) and less xeric sites dominated by chestnut oak (*Q. prinus*), white oak (*Q. alba*), and hickories (*Carya glabra,* and *C. tomentosa*). Occasionally some mesic species will be found. Of this category, the most common are beech (*Fagus grandifolia*), tulip-poplar (*Liriodendron tulipifera*), umbrella tree (*Magnolia fraseri*) and hemlock (*Tsuga canadensis*). Understory species typically are those found in other dry sites such as dogwood (*Cornus florida*), sourwood (*Oxydendrum arboreum*), black locust (*Robinia pseudo-acacia*), persimmon (*Diospyros virginiana*), blackgum (*Nyssa sylvatica*), and serviceberry, (*Amelanchier arborea* var. *arborea*). Another variation of serviceberry, *A. arborea* var. *laevis*, is found mainly on rocky balds in the North Carolina mountains, but has been reported from Oconee County as well.

The shrub layer is well represented in these areas, but because of the nature of the species present, it is not what one would call impenetrable. Principal ones usually encountered are fringe tree (*Chionanthus virginicus*), mountain laurel (*Kalmia latifolia*), lead-plant (*Amorpha fruticosa*), buffalo nut (*Pyrularia pubera*) and several species of ericaceous shrubs, mainly deerberry or squaw-huckleberry (*Vaccinium stamineum*) and mountain blueberry (*V. vacillans*). Of all the above mentioned, the two species of *Vaccinium* seem to be the most evident. American chestnut (*Castanea dentata*) also is evident as small saplings, but once again, it never gets very large.

Along with the relatively open canopy and diffuse shrub layer there is an abundance of herbaceous species. In fact, in places this abundance even rivals that of lower slopes. Common members of the composite family include grass leaved aster (*Heterotheca graminifolia* and *H. nervosa*), pussytoes (*Antennaria* spp.), rosin-weed (*Silphium compositum*), lion's foot (*Prenanthes serpentaria* and *P. altissima*), white-topped aster (*Aster paternus*), and various goldenrods (*Solidago* spp.). Common grasses include oatgrass (*Danthonia* spp.), panic grass (*Panicum* spp.), and broomstraw (*Andropogon scoparius* and *A. virginicus*). There is also a virtual preponderance of wild yam (*Dioscorea villosa*), greenbriers (*Smilax* spp.), and spurge (*Euphorbia corollata*). In some places the ground is nearly blanketed by galax (*Galax aphylla*), with numerous specimens of pipsissewa (*Chimaphila maculata*) and beggar's lice (*Desmodium* spp.) filling in open spots. Other miscellaneous herbs likely to be encountered are cinquefoil (*Potentilla canadensis*), milkweed (*Asclepias variegata*), partridge pea (*Cassia nictitans*) and Indian-physic (*Gillenia trifoliata*). In late summer and early fall, some areas are also dotted with numerous whorled loosestrife (*Lysimachia quadrifolia*), and stems and leaves from the many violets that bloomed earlier.

Ferns are not nearly as common as on the more mesic slopes. Those that do appear are usually bracken fern (*Pteridium aquilinum*), and ebony spleenwort (*Asplenium platyneuron*).

FIGURE 18. Slope exposure effect on vegetation growth. The east-facing slope (right) supports a much more luxuriant growth of mountain laurel (*Kalmia latifolia*) than the west-facing slope (left). In addition there is a less scrubby growth form to arborescent species on the more mesic east-facing slope.

Pine Forests

The upper ridges of the Blue Ridge Province are in sharp contrast to the more mesic slopes and coves. These ridges are decidedly more xeric than the surrounding slopes and often have a very rocky substrate that has been eroded to various extents.

The most evident difference between ridges and their bordering areas is the partial to complete dominance by pines. Several species are evident, with the presence of each species closely associated with fertility and dryness of the soil.

Pine-dominated communities generally occur on three main ridge forms: (1) the continuous and horizontal main ridge separating parallel gorges, (2) the crest of a knob or small hill, and (3) the sloping lead between adjacent coves. For these ridges to have well-developed pine communities, they must be sufficiently elevated above the cove or stream so that they are well exposed. In addition, a south-southwest or southwest exposure is also necessary for pine dominance. In researching pine communities of the Blue Ridge Escarpment, Racine (1966) found that the position and areal extent of pine stands is also related to topography, and that this extent involves the steepness and exposure of the side slopes and the length of the ridge itself. Where a ridge is oriented in an east-west direction, the deciduous forest of chestnut oak may reach the crest on the north slope, whereas on the south slope pine may extend so far down that it borders the cove forest. On a ridge running north-south, the pine vegetation is confined to a narrow strip along the crest.

South Carolina's pine communities are not as well developed as some in the Blue Ridge Escarpment, but there are some that do exemplify these pine-hardwood associations. Most Blue Ridge pine stands have pitch pine (*Pinus rigida*) and scarlet oak (*Quercus coccinea*) as the most common arboreal species. Virginia pine (*P. virginiana*) and shortleaf pine (*P. echinata*) are common associates at elevations under 2800 feet (851.2m). Deciduous species include such common species as sourwood (*Oxydendrum arboreum*), chestnut oak (*Q. prinus*), white oak (*Q. alba*), red maple (*Acer rubrum*), blackgum (*Nyssa sylvatica*), and various hickories. Small un-

FIGURE 19. A chestnut oak-dominated ridgetop forest in Jones Gap. This xeric, rocky ridgetop supports a canopy growth of chestnut oak (*Quercus prinus*), scarlet oak (*Q. coccinea*), and blackgum (*Nyssa sylvatica*), with red maple (*Acer rubrum*), pignut hickory (*Carya glabra*), and black oak (*Q. velutina*) being of secondary importance. Understory is dominated by dogwood (*Cornus florida*), small beeches (*Fagus grandifolia*), sassafras (*Sassafras albidum*), and sourwood (*Oxydendrum arboreum*). Numerous specimens of witch hazel (*Hamamelis virginiana*), buffalo nut (*Pyrularia pubera*), and sweetshrub (*Calycanthus floridus*) occur along with small weak saplings of American chestnut (*Castanea dentata*). Ground cover is usually lacking except for small blueberries and an occasional rattlesnake plantain (*Goodyera pubescens*), dwarf iris (*Iris verna*), or other herbaceous plants.

derstory species and shrubs include sassafras (*Sassafras albidum*), horse-sugar (*Symplocos tinctoria*), and sparkleberry (*Vaccinium arboreum*). Knee-high ericaceous shrubs include deerberry (*V. stamineum*) and huckleberry (*Gaylussacia ursina*). There is usually an abundance of spotted wintergreen (*Chimaphila maculata*) and various species of greenbrier (*Smilax* spp.).

While the above species along with mountain laurel (*Kalmia latifolia*) are usually very prevalent under a closed canopy, an open canopy with a southern exposure tends to produce a still more xeric herbaceous flora. In such situations one tends to find more broomsedge (*Andropogon* spp.), iris (*Iris verna*), bracken fern (*Pteridium aquilinum*), and bird's foot violet (*Viola pedata*). Occurring also are legumes such as false indigo (*Baptisia tinctoria*), pencil flower (*Stylosanthes biflora*) and several species of beggar's lice (*Desmodium* spp.).

In pine communities less xeric than the aforementioned, hardwoods tend to codominate to a greater extent. It is not uncommon in such areas to have a pine-scarlet oak canopy and a nearly closed heath of mountain laurel, huckleberry, and leucothoe (*Leucothoe recurva*). These shrubs are generally associated with broad patches of galax (*Galax aphylla*), numerous specimens of blazing-star (*Chamaelirium luteum*), and cow-wheat (*Melampyrum lineare*).

On dry south-facing slopes the dominants are usually shortleaf pine and various hardwoods such as blackjack oak (*Q. marilandica*), black oak (*Q. velutina*), post oak (*Q. stellata*), southern red oak (*Q. falcata*), and various hickories. In some areas black oak is often present to a relatively large degree, but in other areas it is scarce. Black oak has been found to be particularly succeptible to long periods of very low moisture (Hursh and Haasis, 1931) and gives way to chestnut oaks and shortleaf pine.

FIGURE 20. A rock face and ridge top forest, 2800 feet (851 m) elevation in Jones Gap. A north-facing community (left foreground) is bordered on the top by a chestnut oak-pine ridgetop forest.

Rock Communities

Since the majority of the species and associations found on flat rock communities in the Blue Ridge Province are basically the same as those in the upper piedmont, discussion here will be concentrated on rock face communities, and one should therefore refer to the rock communities section of the piedmont for a discussion of flat rocks.

Near vertical communities can be generally grouped into two divisions; those associated with running water (usually known as spray communities) and those on bare rocks where little running water is present. Spray communities are very hard to characterize as to dominants. Usually the water flow is too swift to allow much growth of flowering plants, although a few species such as river-weed (*Podostemum ceratophyllum*) are capable of attaching to the wet rocks and can avoid being washed away by the current. The majority of herbaceous plants present are mosses and liverworts, although along the outer edges of the spray local herbacous endemics may prevail.

Succession in xeric phases generally takes place on small, narrow ledges where small bits of organic matter and soil accumulate. Small mats of rock spikemoss (*Selaginella tortipila*) are usually the first to form over pioneer species such as *Rhacomitrium heterostichum*. These mats are then invaded by various species of fruticose lichens such as *Cladonia*. Oosting and Anderson (1937) found that in the North Carolina mountains the principal lichen species involved were *C. subcariosa*, with species such as *C. rangiferina, C. tenuis*, and *C. mitis* being subordinates.

Once organic matter has been built up enough to support flowering plant populations, invasion usually begins at the center of the mat and spreads in a centrifugal direction.

FIGURE 21. A spray community. This small waterfall, like others throughout the Blue Ridge Province, has a copious growth of algae, liverworts, and mosses on the sheltered faces of rocks. Additionally, there may be a relatively lush growth of riverweed (*Podostemum ceratophyllum*), a rather coarse flowering plant, only a few inches long, that attaches itself to the wet rocks by small round discs. This is the same plant that covers many rocks and logs in streams and rivers where the water's oxygen content is high.

Pioneers include dwarf dandelion (*Krigia montana*), oatgrass (*Danthonia spicata*), pineweed (*Hypericum gentianoides*), broomsedge (*Andropogon scoparius*), panic grass (*Panicum* spp.), evening primrose (*Oenothera perennis*), coreopsis (*Coreopsis major*), and goldenrod (*Solidago roanensis*). Other species include various briers (*Rubus* spp.), loosestrife (*Lysimachia quadrifolia*), and a couple of ferns such as New York fern (*Thelypteris noveboracensis*) and rockcap fern (*Polypodium virginianum*).

After organic matter has built up to a point where shrubby species can exist, red maple (*Acer rubrum*) and fringe tree (*Chionanthus virginicus*) are found. Occasionally, certain local species of pines can exist if the taproot can find a crack or deep crevice. Naturally these hardy woody species never attain much height except where conditions allow the roots to penetrate into a deep fissure in the rock.

Mesic successional communities usually culminate in species slightly more mesic, although many of the intermediates may be the same as in xeric succession. In addition to various species of *Cladonia*, seepage slopes may also initially develop mats of peat moss (*Sphagnum* spp.) and haircap moss (*Polytrichum* spp.). These species tend to aid in the faster development of an organic layer and the more rapid development of flowering herbaceous plants. In addition to a few individuals of oatgrass and panic grass, numerous specimens of sedges (*Carex* spp.) and saxifrage (*Saxifraga michauxii*) may be found. Where seepage is constantly present, meadow spikemoss (*Selaginella apoda*) may grow, but it is usually limited to sheltered north-facing surfaces. Woody species are generally the same as found in xeric areas, but here again, growth is normally enhanced by the abundance of moisture.

As these mats expand, further invasion of woody species takes place. Earliest invaders include species of briers (*Rubus*

FIGURE 22. Small north-facing vertical rock face communities in Jones Gap. Several small communities have arisen on narrow ledges where tree roots can penetrate cracks and fissures. Many smaller ledges have developed only to the lichen-rock spikemoss stage. The dark vertical streaks on the right are caused by high organic content water seepage from above.

FIGURE 23. A north-facing rock face community, 2800 feet (851 m) elevation. The foreground has an abundance of crustose and fruticose lichens as well as black rock moss (*Grimmia laevigata*) and twisted hair spikemoss (*Selaginella tortipila*) that have provided a base for a variety of grasses and other herbaceous plants. Fringe tree (*Chionanthus virginicus*) and small rhododendrons and blueberries (*Vaccinium* spp.) appear as organic matter increases. Finally, chestnut oaks (*Quercus prinus*) and pines culminate the development of this rock face. The surrounding area is dominated by chestnut oaks and heaths. This rock face community overlooks Jones Gap.

spp.), black chokeberry (*Sorbus melanocarpa*), mountain laurel (*Kalmia latifolia*), purple laurel (*Rhododendron catawbiense*), great laurel (*R. maximum*), and black huckleberry (*Gaylussacia baccata*). In most instances *Rhododendron maximum* is found below 3000 feet (912 m) elevation and *R. catawbiense* at this level and above. As would be expected, there is very little of the latter in South Carolina.

The surrounding forest in most instances is predominately of the chestnut oak-oak association, and in many places it is possible to find a very dense rhododendron thicket bordering the forest. There are usually a few tall chestnut oaks or pines growing among these ericaceous shrubs, and if one is so inclined, one might call the area a chestnut oak-heath community. This growth of shrubs is usually extremely thick and very few species are found underneath. However, one species rather commonly found along the outer edge is downy rattlesnake plantain (*Goodyera pubescens*), but it is also fairly common in the surrounding woodlands.

PART THREE: Piedmont Province

Physiographic Features
    Monadnocks
    Rolling Uplands
    Tributary Stream Valleys
    Major Stream Valleys
    Bluffs
    Flat Rock Outcrops

Natural Vegetation System
    Lentic Water Habitats
    Lotic Water Habitats
    Sandy Stream Beds
    Pond Margins
    Major Floodplain Forests
    Mixed Mesophytic—Cove Hardwoods
    Midslope Oak-Hickory Forests
    "Post Climax" Forests
    Post Oak-Blackjack Oak Ridge Top Forests
    Flat Rock Communities
    Chestnut Oak-Heath Forests
    Old Field Pine Forests

Physiographic Features

MANY YEARS AGO virgin areas of the Piedmont Province were highly fertile and highly productive, as demonstrated by the high degree of agricultural productivity over the last 150 years. However, mismanagement, overcropping, erosion, and a multitude of other factors have reduced the once fertile lands to eroded ridges that require high applications of fertilizers to remain productive.

In South Carolina the piedmont slopes gradually eastward from the foot of the mountains to the fall line, which marks the inner boundary of the coastal plain. Typical piedmont topography is represented by a series of gently rolling areas interrupted by the deeper, steeper valleys of larger streams. The rivers, for the most part, have cut directly across from highland to coastal plain. As we travel across the piedmont from north to south, we pass from hilltop to hillside to valley with only a few level floodplains evident. We also find that there are relatively few sharp breaks in the topography of the lower piedmont except along major river valleys. These ridges and valleys are interwoven by a multitude of small streams that make up the numerous small rivers. The humidity produced by these bodies of water, when coupled with the generally warm, but highly variable range of temperatures, has led to a diversity of plant life. In some respects this diversity may not be as evident as that of the mountain or coastal plain provinces, but in many ways it

is. If we take into consideration the ecological variation from upper piedmont to lower piedmont, combined with the results of plant migrations southward and eastward from the Appalachian extensions and northward and westward from the coastal plain and Florida, we have the potential for a myriad of small disjunct stands of vegetation.

Piedmont forests generally belong to the Oak-Hickory Formation as established by Braun (1950). This classification is well demonstrated by old forest stands that are gradually returning to the oak-hickory dominated status. However, a high degree of habitat diversity in relation to water and soil composition has led to the recognition of several general community types. The most characteristic association is the white oak-black oak-red oak association. Associated species vary widely, with the dominant arborescents being several species of hickory, loblolly and shortleaf pine, black gum and sweetgum. Understory is characterized by saplings of the arborescent stratum as well as by flowering dogwood and sourwood.

FIGURE 24. Glassy Mountain is a typical granite monadnock capped by a thin soil layer supporting a xeric growth of pines and hardwoods. Note the small patches of vegetation on ledges and crevices and the organic water stained vertical sides. Many such monadnocks exist throughout the upper piedmont and even into the sandhill region, although most are much smaller.

Variances in moisture content naturally reflect a characteristic change in vegetational dominants. In more xeric localities, post oaks and blackjack oaks replace red and black oaks, whereas in hydric situations more water-tolerant species abound. Typical forest composition would be dominated by willow oak, swamp chestnut oak and overcup oak, with white oak being of secondary importance.

River tributaries and small streams infrequently subjected to flooding are dominated by beech, ash, hickories, and birch, with willow oaks, redbud, hophornbeam, and musclewood as understory. In addition, there is often a narrow border along the water's edge consisting of willows and alders. Along rivers where alluvial soil has been deposited the characteristic vegetation is very similar to flood plains of the coastal plain, though not as extensive. Domi-

nants are sweet gum, water oak, and white ash, with various pines occasionally intermixed. In addition, tulip-poplar may become dominant on slightly drier sites. Understory and smaller trees characteristically are red maple, boxelder, papaw, and spicebush.

Because of the extreme age of piedmont formations, soils and weathered rocks are of considerable depth; of course there are exceptions. Granite flatrocks occur sporadically more or less throughout, but there is a high concentration near the fall line and again in the northwestern part of the piedmont. Primary succession on these unique areas usually follows one of two lines: it either originates on dry, exposed rock surfaces or in depressions, whether filled with water or not.

The present aspect of piedmont landscape has doubtless come about as a result of one or more erosion cycles. These cycles have left us with an area as complex as anyone would like to make it, yet an area which, for a layman's viewpoint, is relatively unimpressive.

Natural Vegetation System

Freshwater Systems

Ecologists classify freshwater habitats as either lentic (standing water) or lotic (running water). In the piedmont we have a combination and intermingling of the two, which at times makes it very hard to set them apart. There are no sharp boundaries, and depending upon physical and biotic interactions, there may be a shift from one to the other. Erosion and plant growth tend to fill in lakes, thus eventually producing terrestrial habitats, but such forces as high winds, floods, or solution of the underlying substrate may intervene and counteract natural succession. Streams often cut down to

FIGURE 25. Stevens Creek mixed hardwoods, a mixed mesophytic community probably dating from the Pleistocene era and dominated by red oak (*Quercus rubra*), pignut hickory (*Carya glabra*), sugar maple (*Acer saccharum*), and hackberry (*Celtis laevigata*). Its location on a steep north slope of a down-cutting stream probably aids in its perpetuance (top). It is an unusual community in that the soils are basic but the parent material is acidic. One theory for its existence is that the dominants of the Pleistocene became established and returned enough bases to build a basic soil. It is also unusual because it contains some disjunct plant species from both northern and southern distributions, such as the bald cypress (*Taxodium distichum*) along the creek banks (bottom). The combined effects of soil and slope may account for the northern elements, and the circumneutral topsoil and general climate could be responsible for the southern species (*Radford, 1959, 1974*).

FIGURE 26. Piedmont stream and floodplains often have alluvial sandbars that have begun to be invaded by herbaceous plants. The narrow floodplain is dominated by hickories, winged elm (*Ulmus alata*), sycamore (*Platanus occidentalis*), and sweetgum (*Liquidambar styraciflua*). Boxelder maple (*Acer negundo*) and hackberry (*Celtis occidentalis*) are principal understory species. The dense ground cover along the bank is aided in its development by the opening in the canopy over the stream.

the base level as a result of swiftly running water, and when the base level is reached, they remain in a relatively stable condition. In larger streams and rivers, deltas may be deposited and create a shift in the current.

Streams may begin at outlets of ponds and lakes or may arise from springs or seepages in pastures, fields, and woods. Such concentrated flows of ground water range from tiny seep holes, through which the water oozes to form wet spots on the ground, to large fissures in rocks or ground openings that are cleared out by rapidly percolating water. Such springs are usually heterotrophic and have little vegetation differing from the surrounding area. The degree of slope and amount of available water determine the velocity of these small streams that arise from seepage areas and other runoff.

Permanently rooted aquatics are practically nonexistent in swift streams. A combination of water velocity and fluctuations in the water level makes attachment difficult. Instead, filamentous algae such as *Cladophora, Oscillatoria, Spirogyra*, and *Ulothrix* cling to rocks by holdfasts. Also among the filaments of these algae are found many different species of diatoms and desmids. *Phormidium* is another blackish green-colored filamentous algae that becomes even more evident in limestone water. Such extensive stands of algae are very evident to anyone who has ever slipped on green slime-covered rocks and debris while crossing streams. The green alga *Nitella* is often found in sunny stream beds and on the bottoms of ponds and lakes where it often forms extensive subaquatic meadows. It is often confused with vascular plants because of its symmetrically branched growth form, however, its foul-smelling musky odor is a distinctive characteristic.

In slower flowing streams the water is usually deeper and the bottom more or less muddy. In these streams as in others, the velocity of water may vary considerably with the season.

FIGURE 27. This section of the Enoree River in Newberry County is representative of the numerous small rivers of the piedmont. The immediate riverbank forest is composed of numerous species of large trees such as water oak (*Quercus nigra*), ash (*Fraxinus americana*), sycamore (*Platanus occidentalis*), hickory (*Carya* spp.), hackberry (*Celtis occidentalis*), and slippery elm (*Ulmus fulva*). Smaller trees include red maple (*Acer rubrum*) and boxelder maple (*A. negundo*), river birch (*Betula nigra*) and black willow (*Salix nigra*), with characteristic vines such as muscadine (*Vitis rotundifolia*) and poison ivy (*Rhus radicans*) also present. The floodplain is dominated essentially by the same arborescent species with an occasional occurrence of sweetgum (*Liquidambar styraciflua*), American elm (*Ulmus americana*), and basswood (*Tilia heterophylla*). There are few herbaceous species present because of the dense canopy, however, there is an abundance of poison ivy, muscadine, and cane (*Arundinaria gigantea*), especially where there is an opening in the canopy.

Slow-moving streams normally have a high accumulation of aquatic plants and organic matter along the periphery. Such plants as pondweed (*Potamogeton diversifolius*) may extend to the center of the stream. Along stream margins where organic matter is high, one often finds lizard's tail (*Saururus cernuus*) growing in large colonies; in addition, black willow (*Salix nigra*), buttonbush (*Cephalanthus occidentalis*), and alder (*Alnus serrulata*) often form distinctive shrub zones.

As larger streams mature and flow toward the coastal plain, the gradual erosion of stream banks and the less rapid changes in elevation cause more and more streams to become slow moving. With this decrease in velocity, aquatic plants characteristic of pools and ponds replace those normally found in swift streams. But it must be remembered that there is no clearcut boundary between swift- and slow-moving streams; rather, there is a continuum from the mouth to the headwaters.

As opposed to swiftly moving streams and rivers, the still waters of lakes and ponds offer a wider variety of plants and usually a greater abundance. Still bodies of water such as farm ponds and beaver ponds may be temporary or permanent, man-made or natural. They may be large or small and may or may not have an outlet. In general though, we find that shallow, older bodies of water tend to have the most diversified plant life.

In considering the freshwater environment as a whole, the most common plants are algae, with aquatic seed plants ranking second. Of the seed plants, we find that [with the exception of pondweeds (*Potamogeton* spp.) and duckweeds (*Lemna* spp.)] most aquatic higher plants are members of diverse families in which the major species are terrestrial. These rooted aquatics typically form concentric zones within the littoral zone, one group replacing another as the depth of the water increases. Floating aquatics, on the other hand,

may form almost a continuous sheet on the surface at certain seasons, thus "shading out" submerged aquatics.

In discussing the various groups of plants encountered, the usual procedure is to group them as free-floating, submerged-anchored, and emergent. In South Carolina the most commonly reported free-floating plants belong to the duckweed family (Lemnaceae). These plants are very simple flowering plants, and some species are among the smallest flowering plants known. Common duckweed (*Lemna* spp.) is frequently found in the coastal plain but often works its way into the piedmont. Associated with duckweed may be *Spirodela polyrrhiza* and watermeal (*Wolffiella floridana*).

Two free-floating water ferns are becoming more prevalent in South Carolina. Mosquito fern (*Azolla caroliniana*) has scale-like leaves, and is often found growing with duckweed. The leaves may often have a tinge of red or purple, depending upon the season and amount of sunlight they receive. Another member of the same family, water fern (*Salvinia rotundifolia*), is larger and has rounded to oval leaves. The leaves appear to be folded lengthwise and have copious amounts of small bristles. Water fern is a relatively recent addition to our flora, but like so many others, is spreading very rapidly, often covering entire ponds.

Rooted vegetation, as opposed to floating vegetation, is more zonal. There is a distinct progression from shallow water dominated by emergents, to deeper water dominated by anchored plants with floating leaves, and finally to an area dominated by submerged-anchored aquatics.

Cattails (*Typha latifolia*) may be considered characteristic of the outer zone. Other emergents present are various sedges (*Carex* spp.), bulrushes (*Scirpus* spp.), rushes (*Juncus* spp.), and water primrose (*Ludwigia palustris* and *L. alternifolia*). Alligator weed (*Alternanthera philoxeroides*) and slender spikerush (*Eleocharis acicularis*) may be found in large

FIGURE 28. One of a series of beaver dams near Townville, Anderson County, that developed after the U.S. Army Corps of Engineers set aside the area as a wildlife refuge. There are numerous sites with small populations of beavers in the State, but no other quite as large as this.

colonies. While alligator weed is a large plant covering large areas of still water, slender spikerush is a very small, almost filamentous plant inhabiting usually acid, muddy banks where water is only a few inches deep, or it may form extensive mats in deeper water. Bur-reed (*Sparganium americanum*) is often found in the upper piedmont in association with wapato or duck-potato (*Sagittaria latifolia* var. *pubescens*) and pickerelweed (*Pontederia cordata*). The first two are much more widespread, but like many other species, pickerelweed is spreading rapidly.

Buttonbush (*Cephalanthus occidentalis*), black willow (*Salix nigra*), silky dogwood (*Cornus amomum*), and alder (*Alnus serrulata*) are the principal woody species present; however, depending upon the location and surrounding conditions, other species may also exist.

Anchored aquatics with floating leaves are relatively few in number. The most common ones are water lily (*Nymphaea odorata*) and water shield (*Brasenia schreberi*). Cow lily (*Nuphar advena*) may also be present if the water is not too deep. The petiole of this plant is usually strong enough to elevate the leaf blade above the water, but blades may also be found floating.

Submerged anchored aquatics usually have leaves that are thin, finely divided, and adapted for exchange of nutrients and gases with the water. The pondweeds (*Potamogeton* spp.) are most prominent in this zone, and have a wide diversity of leaf shapes. Other submerged aquatics commonly found in one area or another are coontail (*Ceratophyllum demersum*), bushy-pondweed (*Najas quadalupensis*), and waterweed (*Elodea canadensis*). Common *Elodea* is the plant used by aquarium enthusiasts, and it is the same plant that is now causing congestion in areas of the Santee Swamp. Parrot-feather (*Myriophyllum brasiliense*) is very common in the coastal plain and lower to upper piedmont. It is seldom seen flowering but propagates easily by fragmentation.

Along major streams where the current has deposited sandbars or has otherwise established a small floodplain, vegetation tends to be a combination of pond margin species and river floodplain species. Typically, the first shrubby species are black willow (*Salix nigra*) and alder (*Alnus serrulata*). At times, depending upon the extent of the floodplain, these shrubs may form dense thickets. Later, the stand may be invaded by river birch (*Betula nigra*) and sycamore (*Platanus occidentalis*). Typical shrubs are buttonbush (*Cephalanthus occidentalis*), silky dogwood (*Cornus amomum*), elderberry (*Sambucus canadensis*), possum-haw holly (*Ilex decidua*), and two species of the genus *Viburnum*. Wild raisin (*Viburnum cassinoides*) is commonly found along rocky crevices in the upper piedmont and mountains, whereas black haw (*V. rufidulum*) is very widespread throughout the piedmont. One of the most noticeable vines present is virgin's bower (*Clematis virginiana*), which has very beautiful flowers and equally attractive fruits.

Major streams and rivers may be bordered by wide meadows relatively free of shrubs and trees. These areas are dominated by herbaceous species that depend upon strong light and high soil moisture for optimum development. Species of the bluegrass genus (*Poa* spp.), sedges (*Carex* spp.), and bulrushes (*Scirpus* spp.) are the dominant graminoides, while species of the mustard (Brassicaceae), pink (Caryophyllaceae), and buttercup (Ranunculaceae) families tend to make up a large part of the remaining flora.

Common members of the mustard family are winter cress (*Barbarea verna* and *Cardamine hirsuta*), yellow cress (*Rorippa islandica*), peppergrass (*Lepidium virginicum*), and mouse-ear cress (*Arabidopsis thaliana*), which is also very common in lawns and waste places. *Ranunculus* is the principal genus of the buttercup family, and at times the blossoms make a whole field appear yellow. Within the pink family, the principal species are chickweed (*Stellaria media*), mouse-ear

FIGURE 29. A wetland meadow leading to a beaver complex (background). In the extreme background are several large dead trees probably killed when water was backed up by beaver dams. The wetland meadow (foreground) is dominated by alder (*Alnus serrulata*), elderberry (*Sambucus canadensis*), willows (*Salix* spp.), and herbaceous plants such as rushes (*Juncus effusus* and others), barnyard grass (*Echinochloa crusgalli*), jewel-weed (*Impatiens capensis*), tearthumb (*Polygonum sagittatum*), and monkey flower (*Mimulus ringens*). Virgin's bower (*Clematis virginiana*) is also extremely abundant and at some places blankets the other vegetation.

FIGURE 30. An upstream view of the Savannah River where it has been divided by numerous islands. The current is maintained at a constant level at most times by the dam at the Hartwell Reservoir, but the piled up trees and debris on the rocks show that high flood-level water does occur periodically. The immediate floodplain is dominated by ash (*Fraxinus americana*), river birch (*Betula nigra*), winged elm (*Ulmus alata*), and sycamore (*Platanus occidentalis*). Alders (*Alnus serrulata*), musclewood (*Carpinus caroliniana*), red oaks (*Quercus rubra*), and boxelder maple (*Acer negundo*) are in the understory.

chickweed (*Cerastium glomeratum*), and pearlwort (*Sagina decumbens*).

Other easily recognizable species of various families are corn salad (*Valerianella locusta*), geranium (*Geranium carolinianum*), and veronica (*Veronica arvensis* and *V. peregrina*). Also present are henbit (*Lamium amplexicaule*), a very widespread member of the mint family, and various violets, of which wild blue violet (*Viola papilionacea*) and wild pansy (*V. rafinesquii*) are the most common.

Floodplain Forests

Piedmont floodplain communities are very similar in composition to coastal plain floodplain communities, but they are less extensive. The soil is usually of alluvial origin, and in most cases, it is more fertile than the bordering upland soils. Dominant species composition varies slightly depending upon the age of the forest, the width of the floodplain, the steepness of the surrounding land, the composition of surrounding forests, and the presence of disturbing influences. Nevertheless, there are species that are found to be characteristic in such communities. Red gum (*Liquidambar styraciflua*) and tulip-poplar (*Liriodendron tulipifera*) are generally the dominants, with red gum being the earliest dominant and generally occupying wetter sites than tulip-poplar. Codominants associated with both are usually ash (*Fraxinus* spp.), winged elm (*Ulmus alata*), and red maple (*Acer rubrum*). In the upper piedmont there is a greater incidence of river birch (*Betula nigra*) than in the lower piedmont.

In wetter areas dominated by sweetgum, other species appear in greater numbers. Such species include willow oak (*Quercus phellos*), musclewood (*Carpinus carolinianus*), sycamore (*Platanus occidentalis*), and shagbark hickory (*Carya ovata*). Drier sites may have a greater incidence of red oak (*Q.*

FIGURE 31. A well-established growth of riverweed (*Podostemum ceratophyllum*) on an old log and rock in the Savannah River. This plant is very abundant where its substrate consistently remains below water level and oxygen content is high. It is also present in various rivers in the sandhills and on a few waterfalls in the mountains.

FIGURE 32.   A remnant of a once dense stand of cane (*Arundinaria gigantea*) along the Saluda River floodplain in Greenville County. Very few large canebrakes remain today, but in years past they covered acres of bottomlands along piedmont rivers.

rubra), white oak (*Q. alba*), water oak (*Q. nigra*), beech (*Fagus grandifolia*), and mockernut hickory (*Carya tomentosa*).

Shrubs in nearly all cases are dominated by spicebush (*Lindera benzoin*), boxelder maple (*A. negundo*), papaw (*Asimina triloba*), and bladdernut (*Staphylea trifolia*). Also common are deerberry (*Vaccinium stamineum*), maple leaved viburnum (*Viburnum acerifolium*), and strawberry bush (*Euonymous americanus*). In addition, the edges of streams and pools may have a high incidence of willow (*Salix nigra*), alder (*Alnus serrulata*) and buttonbush (*Cephalanthus occidentalis*).

Vines usually found are very common species such as wild yam (*Dioscorea villosa*), greenbrier (*Smilax* spp.), trumpet vine (*Campsis radicans*), Virginia creeper (*Parthenocissus quinquefolia*), wild grape (*Vitis* spp.), and poison ivy (*Rhus radicans*). Honeysuckle (*Lonicera japonica*) and virgin's bower (*Clematis virginiana*) are two of the more showy species present.

The herbaceous layer is in most cases sparse, with most of the species appearing in spring and early summer. Frequently found species include sedges (*Carex* spp.), buttercups (*Ranunculus* spp.), spring beauty (*Claytonia virginica*), jack-in-the-pulpit (*Arisaema triphyllum*), pale corydalis (*Corydalis flavula*), Indian strawberry (*Duchesnea indica*), and birthwort (*Aristolochia serpentaria*).

Outer borders of these floodplains and low-lying mesic sites that are subjected to very infrequent or no flooding generally reach a climax stage with red gum (*Liquidambar sytraciflua*) and tulip-poplar (*Liriodendron tulipifera*) as dominants. This is generally unlike the true floodplain community which eventually results in an oak-hickory climax. Instead, the border community may develop into a post-climax oak-hickory forest.

Species associated with sweetgum and tulip-poplar are usually found to be influxes from both the floodplain and higher ground. Typical species are ash (*Fraxinus* spp.), beech (*Fagus grandifolia*), bitternut hickory (*Carya cordiformis*), river birch (*Betula nigra*), and willow oak (*Quercus phellos*). In addition, redbud (*Cercis canadensis*), hophornbeam (*Ostrya virginiana*), and various other floodplain species may be present.

In the northwestern part of the piedmont there are some unusual combinations of impermeable, basic, clay soils and slight depressions that often result in a semi-boggy habitat. These semi-bog communities contain elements of several community types such as open alluvial woods, river floodplains, and prairies. Canopy dominants are normally the same as in river floodplains. Subcanopy species include persimmon (*Diospyros virginiana*), red cedar (*Juniperus virginiana)* and redbud *(Cercis canandensis)* as well as saplings of dominant canopy species.

Principal shrubs include silky dogwood (*Cornus amomum*), blackhaw (*Viburnum prunifolium*), and bushy forms of briers (*Rubus* spp.). Also found in some localities is Indian currant (*Symphoricarpos orbiculatus*), a shrub with reddish to purplish berries that persist throughout most of the winter. Vines are typified by poison ivy (*Rhus radicans*) and honeysuckle (*Lonicera japonica*).

The herb layer, even though predominant species are sedges (*Carex* sp.) and rushes (*Juneus* sp.), is possibly the most interesting in the community. There is normally a noticeable zonation from boggy to drier areas, with species of wild garlic (*Allium bivalve* and *A. vineale*), spring beauty (*Claytonia virginica*), bedstraw (*Galium* spp.), buttercups (*Ranunculus* spp.), and violets (*Viola palmata* and others). In addition, wild Easter lily or atamasco lily (*Zephyranthes atamasco*) occurs occasionally. There is one prairie disjunct, wild hyacinth (*Camassia scilloides*), which grows in such a semi-bog community in York County.

These hardwood semi-bog communities are usually of a climax nature, and will return after a disturbance, but the reoccurrence of disjuncts such as *Camassia* is unlikely since these plants were possibly "trapped" in these unique localities during post-glacial migrations (Radford, 1974).

Mixed Mesophytic–Cove Hardwood Forests

Under ideal conditions of moisture and exposure, such as those often found on relatively cool, moist, well-drained north-facing or northeast-facing slopes, mixed mesophytic-cove hardwood forests are often found. These forests are best developed on steep slopes that are being actively cut in the present erosion cycle. Typical piedmont sites include talus slopes, river bluffs, and deep coves that are actively being cut by a stream. It is here that we find the greatest diversity of species in the piedmont, such as species from river floodplains and low elevations. Dominants in nearly all cases, along with tulip-poplar (*Liriodendron tulipifera*), are beech (*Fagus grandifolia*), red oak (*Quercus rubra*), white oak

FIGURE 33. A mixed mesophytic community on a northeast-facing piedmont slope. There is an abundance of large beech (*Fagus grandifolia*) and white and red oaks (*Quercus alba* and *Q. rubra*). Several small umbrella trees (*Magnolia fraseri*) are present along with numerous sweetshrubs (*Calycanthus floridus*) and a large colony of *Nestronia umbellula*. There is also an abundance of dogwood (*Cornus florida*), redbud (*Cercis canadensis*), hophornbeam (*Ostrya virginiana*), and buckeye (*Aesculus sylvatica*).

(*Q. alba*), willow oak (*Q. phellos*), pignut hickory (*Carya ovata*), bitternut hickory (*C. cordiformis*), white ash (*Fraxinus americana*), and two maples (*Acer rubrum* and *A. saccharum* spp. *floridanum*). Species usually found in the understory, in addition to saplings of the dominants, are hophornbeam (*Ostrya virginiana*), slippery elm (*Ulmus fulva*), dogwood (*Cornus florida*), redbud (*Cercis canadensis*), and buckeye (*Aesculus sylvatica*).

In addition, many species from other localities may also be represented in large numbers. Chief among these are mountain laurel (*Kalmia latifolia*) and storax (*Styrax grandifolia*). Another species of storax (*S. americana*) is common in swamp forests, alluvial woods, and on stream banks of the inner coastal plain. Two other genera are often well represented: the genus *Rhododendron* is represented by wild azalea (*R. nudiflorum*) and Carolina laurel (*R. minus*), and *Magnolia* by umbrella tree (*M. tripetala*) and cucumber tree (*M. acuminata*).

The shrub layer is usually very open, allowing the development of numerous different herb associations. Of the shrubs that are present, papaw (*Asimina triloba*), spicebush (*Lindera benzoin*), and sweetshrub (*Calycanthus floridus*) are very representative. Vines usually common include muscadine (*Vitis rotundifolia*), crossvine (*Anisostichus capreolata*), poison ivy (*Rhus radicans*), and greenbrier (*Smilax* spp.).

Ferns commonly present are Christmas fern (*Polystichum acrostichoides*), ebony spleenwort (*Asplenium platyneuron*), marginal shield fern (*Dryopteris marginalis*), and southern lady fern (*Athyrium asplenioides*). One may often find relatively large colonies of a mountain disjunct, maiden-hair fern (*Adiantum pedatum*), in scattered localities.

Of the common herbaceous species found, the Liliaceae is very well represented, especially in the spring. Bellwort (*Uvularia perfoliata*), trout lily (*Erythronium americanum*),

fly-poison (*Amianthium muscaetoxicum*), blazing-star (*Chamaelirium luteum*), Solomon's seal (*Polygonatum biflorum*), and false Solomon's seal (*Smilacina racemosa*) are also very common. Other members of the lily family present may be trillium (*Trillium cuneatum* and *T. catesbaei*), and of course, greenbrier (*Smilax* spp.). The Amaryllidaceae is also well represented by yellow star-grass (*Hypoxis hirsuta*). Other monocots typically found are dwarf iris (*Iris cristata*), wild yam (*Dioscorea villosa*), downy rattlesnake plantain (*Goodyera pubescens*), crane-fly orchid (*Tipularia discolor*), and jack-in-the-pulpit (*Arisaema triphyllum*).

Herbaceous dicots are also very prominent in the spring, with some persisting well into summer and autumn. Among the most widely seen are heart leaf (*Hexastylis arifolia* and *H. virginica*), chickweed (*Stellaria pubera*), avens (*Geum canadensis*), two species of bedstraw (*Galium circaezans* and *G. pilosum*), and spring beauty (*Claytonia virginica*). Others often encountered are wood sorrel (*Oxalis stricta*), wild geranium (*Geranium carolinianum*), black cohosh (*Cimicifuga racemosa*), and windflower (*Thalictrum thalictroides*). Along stream banks or on sheltered slopes we can also find mayapple (*Podophyllum peltatum*) in large colonies, as well as foam flower (*Tiarella cordifolia*) and hepatica (*Hepatica* spp.).

Midslope Forests

Midslope oak-hickory forests occupy a wide variety of sites. They may be described as a continuum of submesic to mesic, well-drained soils codominated by white oak (*Quercus alba*), black oak (*Q. velutina*), red oak (*Q. rubra*), and in some instances Spanish oak or southern red oak (*Q. falcata*).

While white oak is pretty well ubiquitous, the others are more or less limited to certain ranges by soil moisture content. Red oak occupies the most fertile and most mesic sites, which correspond to lower slopes. Black oaks, on the other

FIGURE 34. A typical midslope oak-hickory community dominated by red oak (*Quercus rubra*), white oak (*Q. alba*), shagbark hickory (*Carya ovata*), pignut hickory (*C. glabra*), and red maple (*Acer rubrum*). The understory is principally canopy saplings, dogwoods, and musclewoods (*Carpinus caroliniana*), with an occasional holly (*Ilex opaca*) and small beech (*Fagus grandifolia*).

hand, predominate on higher and drier sites. The driest of the midslope regions is dominated in scattered localities by southern red oak (*Q. falcata* var. *falcata*). Another variation of this species (*Q. falcata* var. *pagodaefolia*) is found principally in low areas of the coastal plain, but it may also be found in low areas of the piedmont.

Associated arborescents are blackgum (*Nyssa sylvatica*), post oak (*Q. stellata*), red maple (*Acer rubrum*), and various species of hickory (*Carya* spp.). Tulip-poplar (*Liriodendron tulipifera*) is also a principal associated species on more mesic sites. Saplings of this species dominate the understory along with numerous dogwoods (*Cornus florida*) and sourwood (*Oxydendrum arboreum*). Red cedar (*Juniperus virginiana*) and various pines (*Pinus echinata, P. taeda*, etc.) may be present in younger stands, but gymnosperm reproduction is decreased in older stands once a hardwood canopy is developed.

FIGURE 35. The winter aspect of a small piedmont stream floodplain and bluff. Such oak-hickory dominated areas support a tremendous abundance and variation of spring annuals (*courtesy S.C.W.M.R. Department*).

Shrubby species that grow large enough to be considered part of the understory include serviceberry (*Amelanchier canadensis*), holly (*Ilex decidua*), sassafras (*Sassafras albidum*), various species of *Viburnum*, including blackhaw (*V. prunifolium*), and persimmon (*Diospyros virginiana*). Also established may be fringe tree (*Chionanthus virginicus*).

Most areas of the oak-hickory association will include many of the same species of vines. Among these will be Virginia creeper (*Parthenocissus quinquefolia*), several species of greenbrier (*Smilax bona-nox, S. glauca, S. rotundifolia*), wild grapes (*Vitis* spp.), Japanese honeysuckle (*Lonicera japonica*), trumpet vine (*Campsis radicans*), and briers (*Rubus* spp.). In certain areas coral honeysuckle (*Lonicera sempervirens*) may be abundant.

Ferns usually found in these midslope associations are ebony spleenwort (*Asplenium platyneuron*), Christmas fern (*Polystichum acrostichoides*), and the epiphytic common polypody (*Polypodium virginianum*). The more mesic areas have rattlesnake fern (*Botrychium virginianum* and *B. biternatum*) with bracken fern (*Pteridium aquilinum*) being found in submesic habitats.

Many different shrub-herb communities exist in association with various site conditions; however, several species are rather widespread. These include beggar's lice (*Desmodium* spp.), bedstraw (*Galium circaezans*), spotted wintergreen (*Chimaphila maculata*), partridge berry (*Mitchella repens*), hawkweed (*Hieracium venosum*), green and gold (*Chrysogonum virginianum*), and elephant's foot (*Elephantopus carolinianus*). Two orchids, downy rattlesnake plantain (*Goodyera pubescens*) and crane-fly orchid (*Tipularia discolor*), are fairly abundant, the former being more abundant in the upper piedmont. The lily family is also well represented by Catesby's trillium (*Trillium catesbaei*) and false Solomon's seal (*Smilacina racemosa*), along with blazing-star or devil's bit

(*Chamaelirium luteum*). Also very common are spring iris (*Iris verna*) and Indian pink (*Spigelia marilandica*).

Along small creeks and on north-facing slopes one may find more diversified flora, but in general, the larger dominants are the same. Mesic low-lying slopes and north-facing or east-facing slopes with optimally fertile soils may develop further than the more typical upland oak-hickory climax forest. In these "post-climax" areas and in narrow coves or ravines, many of the herbs, as well as many of the arborescents, are the same ones found in mixed-mesophytic cove hardwood communities. Dominant trees include sweet-gum (*Liquidambar styraciflua*), shagbark hickory (*Carya ovata*), and several oaks. These oaks most frequently found are willow oak (*Quercus phellos*), red oak (*Q. rubra*), white oak (*Q. alba*), and overcup oak (*Q. lyrata*). The best key indicator species, in addition to those found on mid-slope areas, is probably southern sugar maple (*Acer saccharum* spp. *floridanum*).

On upper slopes, ridgetops, and similar drier sites, a vegetation pattern similar to the blackjack oak-post oak pattern exists, but also includes species found in more mesic areas. Soils in these areas are usually severely eroded and normally have poor water-holding capacity. This normally results in post oak (*Q. stellata*) and white oak (*Q. alba*) as codominants, with post oak not being as scrubby as in the post oak-blackjack oak forest.

Associated with white and post oak are the more mesic oaks and hickories, blackgum (*Nyssa sylvatica*), and red maple (*Acer rubrum*). Pignut hickory (*Carya glabra*) and mockernut hickory (*C. tomentosa*) are the principal hickories present, with black oak (*Q. velutina*) and chestnut oak (*Q. prinus*) the principal associated oaks. Frequently scarlet oak (*Q. coccinea*) may be found, but it increases in frequency in the upper piedmont on drier ridges. Shortleaf pine (*P. echinata*) and loblolly pine (*P. taeda*) are the most frequently encountered pines, with Virginia or scrub pine (*P. virginiana*) also increasing toward the upper piedmont.

Understory is characteristically composed of dogwood, red maple, and sourwood (*Oxydendrum arboreum*). The shrub layer is dominated by ericaceous plants such as rhododendrons, blueberries, and huckleberries, and by holly (*Ilex opaca*) and blackhaw (*Viburnum prunifolium*). Vines and herbaceous plants are similar in species composition to that of scrub oak habitats, although a few more mesic herbs are present such as wild ginger (*Hexastylis arifolia*) and cinquefoil (*Potentilla canadensis*).

Ridgetop Forests

The most xeric forest communities in the piedmont occupy the driest ridgetops, south-facing and west-facing slopes and bluffs, and upland sites with thin soils or with an underlying clay subsoil. The basic community type in these areas is the post oak-blackjack oak forest. Communities of this sort usually have a relatively open canopy and a mixed, but rather sparse, shrub layer predominated by heaths.

The dominant gymnosperm is shortleaf pine (*Pinus echinata*), although in some instances red cedar may appear in great numbers. Blackjack oak (*Q. marilandica*) and post oak (*Q. stellata*) are the principal large trees, with other oaks represented by white oak (*Q. alba*), red oak (*Q. rubra*), and southern red oak (*Q. falcata*). Other trees include persimmon (*Diospyros virginiana*), black gum (*Nyssa sylvatica*), mockernut hickory (*Carya tomentosa*), and pale hickory (*C. pallida*). The understory is very sparse, but where present may include dogwood (*Cornus florida*), redbud (*Cercis canadensis*), and wild black cherry (*Prunus serotina*), as well as saplings of the arborescent species.

Shrubs belong mostly to the Ericaceae, with *Vaccinium*

being the most frequently encountered genus. Principal species include deerberry (*V. stamineum*), low blueberry (*V. vacillans*), and an even shorter species, *V. tenellum*. The last species is usually the smallest of the three with *V. vacillans* being intermediate in height and frequently forming extensive colonies on steep dry slopes. Spotted wintergreen ((*Chimphila maculata*) is quite characteristic of these forests, and although it rarely attains a height over 4 to 5 inches (10.2 to 12.7 cm), its woody habit still classifies it as a shrub. There are very few vines represented in this association, but where present, muscadine (*Vitis rotundifolia*) and Japanese honeysuckle (*Lonicera japonica*) are characteristic.

The herbaceous layer is usually sparse, and also has xeric species predominating. Greenbrier (*Smilax bona-nox*), and blackberry (*Rubus argutus*) are most noticeable because of their thorny stems. On the other hand, beggar's lice (*Desmodium* spp.) is equally noticeable. Other species present may be bedstraw (*Galium pilosum*) and several legumes such as pencil flower (*Stylosanthes biflora*) and lespedeza (*Lespedeza* spp.). Common grasses are broomstraw (*Andropogon scoparius*) and panic grass (*Panicum* spp.).

One variation of the post oak-blackjack oak forest type is found in areas with abundant rock outcrops and/or on steep south-facing slopes. The dominant arborescents are the same as in post oak-blackjack oak areas, but with more of a scrubby growth form. The heaths and weedy herbs and grasses are still prevalent, but because of the dry rocky habitat, unique herbs may also develop.

Sedges such as *Bulbostylis capillaris* and *Cyperus filiculmis* may become very evident in these areas, but the addition of certain ferns makes up the most noticeable difference. While ebony spleenwort (*Asplenium platyneuron*) is present on these sites and is also very widespread over the State, blunt lobed woodsia (*Woodsia obtusa*), wooly lip-fern (*Cheilanthes tomentosa*), and hairy lip-fern (*C. lanosa*) characterize such rocky places.

FIGURE 36. A post oak-blackjack oak forest on the southwestern slope of a piedmont ridgetop on Parson's Mountain in Newberry County. Dominants are post oak (*Quercus stellata*), blackjack oak (*Q. marilandica*), and white oak (*Q. alba*), with an occasional shortleaf pine (*Pinus echinata*) and red cedar (*Juniperus virginiana*). The understory is characterized by species such as sourwood (*Oxydendrum arboreum*), persimmon (*Diospyros virginiana*), and dogwood (*Cornus florida*), whereas the shrub layer is principally ericaceous in nature.

Flat Rock Communities

Flat rock communities may best be described as a study in succession. In typical flat rock communities of the piedmont, there are large expanses of bare rock covered only by crustose lichens. Gradually, as organic material and moisture increase, there is a characteristic change in vegetation to foliose lichens and mosses, then to certain species of herbaceous plants, and finally to a xeric herb-shrub community. Succession may also start in water-filled depressions, following general hydric stages. Although succession is most commonly viewed in mats, simultaneous growth takes place from the fringes of the bare rock followed by a transition zone, which progresses to the dominant vegetation of the region.

The bare rock stage of the community may be dominated by crustose lichens such as *Verrucaria* spp. Very few other species are able to survive such extreme conditions for any length of time. But as time progresses and a thin layer of organic matter develops, a moss-lichen mat community gradually evolves. Dominants include *Grimmia laevigata, Camplyopus flexuosus*, and *Dicranum scoparium*. The center of these areas has the highest organic and moisture content and, as a result, is invaded by several species of reindeer moss lichens (*Cladonia* spp.). Most common species are *C. caroliniana, C. rangiferina, C. subtenuis*, and *C. arbuscula*. Several species of *Parmelia* are also common, such as *P. hypotropa, P. caperata*, and *P. michauxiana*. In localized areas *Usnea* may also be evident (Burbanck and Platt, 1964; McVaugh, 1943).

Herbaceous species are present in nearly all localities, with the amount of soil, moisture, and organic matter being the chief determinants of which species is present and its relative abundance. Herbs most commonly found have shallow, fibrous root systems and are very efficient in obtaining moisture from dew and any other source. Such species as pineweed (*Hypericum gentianoides*), tufted-sedge (*Bulbostylis capillaris*), rock sandwort (*Arenaria brevifolia* and *A. groenlandica*). and toadflax (*Linaria canadensis*) may usually be found, although they are usually not present in the same area. As conditions allow, other small annuals may be found. *Arenaria uniflora* is one such annual found only in a few counties of the lower piedmont and one county in the upper piedmont.

One variation of this community lies in very shallow depressions. These depressions may contain enough moisture to allow a dominance of stonecrop (*Sedum smallii*). Associated herbs in relatively large quantities are rock sandwort (*Arenaria brevifolia*), various grasses such as bentgrass (*Agrostis hyemalis* and *A. elliottiana*), sedges such as *Cyperus aristatus*, beak rush (*Rhynchospora globularis*), and several species of *Carex*. Often haircap moss (*Polytrichum ohioensis*) may invade the area, but it does not become dominant unless moisture conditions improve drastically.

In areas where *Polytrichum* is dominant, other mosses have usually been eliminated by competition and in their places have developed various graminoids such as broomsedge (*Andropogon* spp.), rush (*Juncus georgianus*, a species endemic to such areas), sedges (*Carex* spp. and *Rhynchospora* spp.). Various forbs such as false pimpernel (*Lindernia monticola*), sheep-sorrel or sour-grass (*Rumex acetosella*), tearthumb (*Polygonum sagittatum*), and knotweed (*Polygonum pensylvanicum*) are found as well as composites such as coreopsis (*Coreopsis grandiflora*), elephant's foot (*Elephantopus tomentosus*), squaw-weed (*Senecio tomentosus*, another flat rock endemic), and certain species of goldenrod such as *Solidago erecta* and *S. altissima*. (*S. erecta* is usually more common in the mountain counties and upper piedmont).

FIGURE 37. Large rock outcrops such as the one at 40 Acre Rock in Lancaster County are particularly unfavorable habitats for plant colonization because of long dry summers and high temperatures. Specially adapted plants, including several rock endemics, grow here in spite of the extremely adverse environmental conditions. These granite rocks (top left) are covered by a lush growth of black rock moss (*Grimmia laevigata*) as well as crustose and foliose lichens. The depression in the center has accumulated enough organic matter and soil particles to support a growth of flowering plants such as rockwort (*Arenaria uniflora*), stonecrop (*Sedum smallii*), and various grasses. A small semipermanent (top right) pool at the edge of an "island of succession." The outer edge of the vegetation mat (foreground) supports a thick growth of rockwort. In the center of the mat are numerous herbaceous species as well as small junipers (*Juniperus virginiana*), sweetgums (*Liquidambar styraciflua*), and winged elm (*Ulmus alata*). This pool supports an abundant growth of two rock endemics (bottom left). The fine grass-like plants growing in the water are quillwort (*Isoetes melanopoda*) and at the rear of the mat are two large groups of cottony groundsel (*Senecio tomentosus*), a yellow flowering composite with very fuzzy leaves. Damp depressions represent the best development on the rocks (bottom right). The stages of development are essentially the same as on other places on the rock, but the additional moisture results in a lusher and more rapid growth, with hardwoods such as these appearing in the later stages.

Where conditions permit, certain woody plants such as winged sumac (*Rhus copallina*), red cedar (*Juniperus virginiana*), sparkleberry (*Vaccinium arboreum*), and deerberry (*V. stamineum*) are found. Neither of the latter two species reaches its full height because of limiting soil conditions.

Hydric successional stages follow much the same pattern of development once *Polytrichum* becomes dominant. However, several stages leading up to this are quite different. Large, shallow rock depressions containing standing water usually have a thin layer of soil washed in by rains. Outcrop endemics such as quillwort (*Isoetes melanopoda*), and flame flower (*Talinum teretifolium*) are the principal species present at first, with others appearing later. As the depression fills in, various mosses such as *Polytrichum* and *Sphagnum* may appear, with other moss species such as *Hedwigia ciliata* and *Bryoandersonia illecrebra* appearing later. The last three also occur on seepage paths. During long droughts, shallow depressions may dry up, and further succession will be prevented.

Successional stages for rock outcrops usually proceed from the edge of a developing mat of vegetation to the center in definitely noticeable concentric rings. However, advanced stages of mats developing from hydric stages may be longer in maturing because of the semi-aquatic environment. These same mats, once established, tend to exist longer because of the available moisture.

Normally one would expect the outcrops to eventually be covered by soil and vegetated by plants characteristic of the adjacent community, but this is not always true as demonstrated by the relative stability of rock outcrops and the slowness of their xeric primary succession. The slow weathering of granite and other outcrop substrates contributes to a very slow development of vegetation mats, and coupled with slow changes and destruction of these mats by

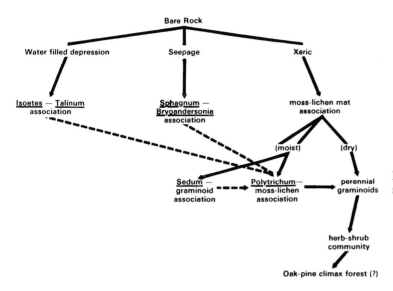

FIGURE 38. Possible successional stages of flat rock communities in the piedmont.

mechanical means (i.e., hard rains, animal life, etc.), tends to result in a relatively stable condition. Conditions found on piedmont rock outcrops are rarely duplicated elsewhere.

It should also be pointed out that there is a relatively large number of endemics that have developed on these rocks; and also, though relatively stable, these communities are very susceptible to human activities and fire during dry periods. The entire soil mat of a thousand years of succession can easily be burned away in a single fire.

Chestnut Oak-Heath Forests

Chestnut Oak (*Quercus prinus*) becomes the dominant quercine species found on subxeric sites with gentle west-facing or south-facing slopes. Although occurrences of this forest type are principally found in the upper piedmont and mountains, areas surrounding large granite outcrops also exhibit a chestnut oak tree dominance and a heath-dominated shrub layer. Soils in these areas are acid, accounting in part for the absence of red cedar (*Juniperus virginiana*) and other plants usually found on neutral to basic soils.

Chestnut oak is the principal tree indicator species, and it is best developed on the more xeric slopes. On more protected northeasterly slopes, red oak (*Q. rubra*) is present in greater numbers because of optimum moisture conditions. Other tree species include oaks such as white oak (*Q. alba*), black oak (*Q. velutina*), blackjack oak (*Q. marilandica*), and post oak (*Q. stellata*), and hickories such as mockernut (*Carya tomentosa*) and pignut (*C. glabra*). Pines normally are absent except in open areas, and then shortleaf (*Pinus echinata*) and Virginia pine (*P. virginiana*) are the principal ones present. The diversity of subcanopy species is restricted. In addition to transgressives, red maple (*Acer rubrum*), dogwood (*Cornus florida*), blackgum (*Nyssa sylvatica*), sourwood (*Oxydendrum arboreum*), and sassafras (*Sassafras albidum*) predominate.

Lianas (or vines) are the common ones found everywhere: muscadine (*Vitis rotundifolia*), poison ivy (*Rhus radicans*), greenbrier (*Smilax* spp.), cross-vine (*Anisostichus capreolata*), and yellow jessamine (*Gelsemium sempervirens*). Herbaceous species very with local soil conditions, but shrub layer composition is fairly constant insofar as the ericaceous genera tend to predominate. The genus *Vaccinium* is best represented by sparkleberry (*V. arboreum*), deerberry (*V. stamineum*), and low blueberry (*V. vacillans*) in the upper piedmont. *V. vacillans* is replaced in importance by *V. tenellum* in the lower piedmont. Other species present are dwarf huckleberry (*Gaylussacia dumosa*), blackhaw (*Viburnum prunifolium*) and spotted wintergreen (*Chimaphila maculata*).

Old Field Pine Forests

Because of numerous abandoned fields throughout the piedmont, old field succession is particularly noticeable, and subclimax pine stands are very conspicuous. In South Carolina loblolly pine (*Pinus taeda*) and shortleaf pine (*P. echinata*) predominate, but may give way to dense stands of Virginia pine (*P. virginiana*) in the upper piedmont.

Before actual invasion by pines, light-textured soils are usually invaded by a community dominated by relatively shallow-rooted herbs. Initial herb invaders vary greatly depending upon local moisture holding capacity and fertility of the soil. On drier sites crabgrass (*Digitaria sanguinalis*) and fleabane (*Erigeron canadensis* and *E. strigosus*) are early dominants. These are replaced in importance during the second and third years by various asters (*Aster* spp.) and broomsedge (*Andropogon* spp.). *Andropogon* may begin dominance earlier in heavy soils, but *Digitaria* usually precedes *Andropogon* because of its drought resistance (Keever, 1950). On severely eroded soils, initial dominants may be poor-Joe (*Diodia teres*) and three awn grass (*Aristida* spp.), with plantain (*Plantago aristata*) appearing somewhat later.

Associated species during the initial two years typically are pearlwort (*Sagina decumbens*), evening primrose (*Oenothera laciniata*), plantain (*Plantago virginica*), dwarf dandelion (*Krigia virginica*), polypremum (*Polypremum procumbens*), Venus' looking glass (*Specularia perfoliata*), field garlic (*Allium vineale*), mouse-ear cress (*Arabidopsis thaliana*), and whitlow-grass (*Draba verna*). Asters later lose dominance because their seedlings are relatively intolerant of shade (Keever, 1950).

Initial pine invasion usually begins during the third year with a closed stand forming in approximately 10 to 15 years. The first major occurrence of hardwoods usually takes place about 20 years after abandonment, with a 40-year-old pine stand normally showing a hardwood understory. In badly eroded areas or where the pines are not so dense, sourwood (*Oxydendrum arboreum*), persimmon (*Diospyros virginiana*), and red cedar (*Juniperus virginiana*) may be co-invaders

FIGURE 39. A dense stand of yellow pine (*Pinus virginiana*) in upper Greenville County. Understory and herbaceous growth are almost completely absent.

with pines (Billings, 1938; Bormann, 1953).

Initial hardwood invasion in most cases is by species having easily transported fruits and wide ecological tolerances such as maples and sweetgums. Further invasion is by oaks and hickories. Stands over 75 years old usually have aging pines replaced by hardwoods and by 150 to 200 years the stand is dominated by oaks and hickories. Shrub and herbaceous layers usually develop slowly and finally attain characteristic growth typical of the dominant community (Bormann, 1953).

Hodgkins (1958) reports that fires cause increased rates of mineralization and nitrification of soil organic matter and also relieve the seedbed of a smothering mantle of litter. In addition, he reports that after a fire, the forbs are the first to respond, particularly legumes and composites; however, vegetation from living rootstocks of grasses, vines, shrubs, and herbs more than assert themselves over approximately three years.

PART FOUR Sandhill Province

Physiographic Features
      Fall Line Sandhills
      Fall Line Rock Outcrops
      Red Hills
      Undrained Shallow Depressions

Natural Vegetation System
      Turkey Oak Barrens
      Scrub Oak Barrens
      Xeric Pine-Mixed Hardwoods
      Pocosins
      Bay Forests
      Hillside Bogs

Physiographic Features

T HE FALL LINE SANDHILLS of South Carolina lie in a discontinuous belt some 5 to 15 miles (8.1 to 24.2 km) wide through the center of the midlands, roughly paralleling the coastline. This fall line area is sculpted by the vigorous erosion of streams passing from the hard crystalline bedrocks of the piedmont to the loose soft sands of the coastal plain. The streams undergo rapid descent and often form shoals in major rivers such as those formed in the Broad and Saluda rivers near Columbia. At other places small streams form waterfalls, such as at Peachtree Rock near Edmund in Lexington County. Also, at some places in Edgefield and Lexington counties, the edge of the coastal plain has worn away enough to produce steep bluffs overlooking valleys, although some wind shifting has modified the dune structure (Cooke, 1936).

For many years formation of the sandhills has been discussed, with the most popular theory being that they are the remnants of former beaches of the Cretaceous period some 130 million years ago. However, Heron (1958) feels that vigorous erosion of the Appalachian Chain resulted in the deposit of a blanket of coarse particles known as the Tuscaloosa formation which now extends well past our present shoreline. On top of the Tuscaloosa formation in some places is an unfossiliferous, non-marine sand and gravel composing the Citronelle formation (Berry, 1914, 1916) to which the

FIGURE 40. One of the many exposed rocks along the fall line near Edmund in Lexington County. The rock is cross-bedded sandstone cemented by impregnations of hematite, a reddish-colored iron oxide. The lower layers of the rock contain fossils dating to the upper Cretaceous and/or lower Tertiary (*rock evaluated by G. L. Stirewalt*).

FIGURE 41. The Broad River at Columbia. As the river passes through the fall line, numerous crystalline rocks are exposed and create rapids where the elevation drops rapidly. Here the river is relatively flat, creating a unique hydric habitat dominated by black willows (*Salix nigra*) and alders (*Alnus serrulata*), with an occasional ash (*Fraxinus*). The extensive herbaceous mats are water willow (*Justicia americana*), which is very common in rocky and sandy riverbeds.

sandhills belong. Since the Citronelle formation only locally covers the Tuscaloosa formation in areas of fluvial fans, fall line sandhills are believed to be alluvial deposits derived from the Tuscaloosa formation, and as a result, have probably been available for plant colonization since the Cretaceous period (Duke, 1961).

More recent Eocene deposits of the coastal plain are common in southeastern South Carolina and extend well into Georgia where they may result in reddish soils of the Orangeburg Type. During the Oligocene period little of South Carolina was inundated. The Miocene epoch encompassed three invasions by seawater, but none resulted in any deposits near the fall line. The Pliocene also resulted in no sandhill deposits. So apparently events in these latter two periods had little effect on the Sandhill Province. At many locations the fall line sandhills are higher than the adjacent piedmont. Some explanations of this phenomenon are the following: (1) the clays of the piedmont are subject to sheet erosion, a process that does not occur in extremely porous sandhill soils; (2) the fall line sandhills are subject to regional uplift (Fenneman, 1938); (3) the dunes are possibly wind deposits (Flint, 1947); and/or (4) some other method of sandhill formation may have been involved.

It therefore seems that the sandhills, as presently understood, were not formed by coastal dunes in the Pleistocene, nor did they receive any marine sediments in the Pleistocene. They are therefore best regarded as discontinuous weathered remnants of the continental phase of the Tuscaloosa formation (Duke, 1961) which dates back to the Mesozoic era.

While the flora of the Carolina sandhills is in itself unique, it is not limited to the region of the sandhills. This xerophytic type of flora has arisen wherever conditions have been suitable. Such conditions include, naturally, an extremely permeable layer of sandy soil that is leached of nutrients by rapidly percolating water. In some instances a clay subsoil may be present, resulting in a characteristic change in dominant plants. These sandhill soils have low water-holding capacity (except where a clay subsoil is present), and the depth of the water table plays an extremely important part in influencing habitats and successional patterns. In fact, the complete transition from a xeric turkey oak barren to a hydric bay or pocosin can occur within a remarkably short distance, often with very little ecotone.

In an overview of Sandhill Province communities, several things are very evident:

1. Water seems to be the most noticeable variant;
2. Sterility of the soil varies considerably;
3. High insolation is limiting in certain areas;
4. Fire is an important natural factor; and
5. The presence of a clay subsoil is more conducive to floristic variation.

Consider the importance of elevation above the water table as the predominant ecological determinant. Very basically, turkey oak barrens are most xeric and scrub oak barrens less xeric, with a steady increase of moisture to hydrophytic bogs and swamp forests. Considering that a typical succession will lead to the most mesophytic flora possible for the region, one would expect changes from turkey oak barrens to scrub oak barrens, the latter being more mesic. However, with the frequency of fires in the sandhill region, the sterility of the soil in general, the high amount of insolation, and the relatively small number of species ecologically tolerant of such conditions, succession seems to be nearly nonexistent, at least where the water table remains well below the surface. Generally the only noticeable change in turkey oak barrens is (1) a gradual increase of longleaf pine where fire is a frequent factor, and (2) an elimination of longleaf pine as a codominant in areas protected from fire (Wells, 1924, 1928, 1967; Wells and Shunk, 1931).

In areas where the water table approaches and/or intersects the surface, conditions are much more conducive for succession to occur. If a clay subsoil is present, the resulting vegetation varies from the quite mesic sandhill type to hydric types often characterized by alder (*Alnus serrulata*). Areas of this type are usually more protected from fire and, as a result, are more conducive to succession to mesic hardwoods. Hillside bogs of this type are usually not well defined and may be construed as an extended border of pocosin communities where there is no clear separation to more xeric species.

Hydric ecosystems of the Sandhill Province are controlled largely by the amount and periodicity of flooding and the resulting drainage patterns. Bogs arise where the water table usually intersects the surface only in the spring and is below the surface most of the remainder of the year. Drainage usually is poor, and organic materials accumulate. Mosses such as *Sphagnum* form extensive covers of floating mats over a mushy bottom that contains low oncentrations of potassium, nitrogen, and oxygen (Smith, 1966). Unfavorable conditions for most plants are compounded by the strongly acidic environment. Pioneer species in these areas are usually acidophyllous ericads. The resulting acid condition also tends to restrict subclimax dominants to needle-leaved species such as pond pine (*Pinus serotina*) and Atlantic white cedar (*Chamaecyparis thyoides*), depending upon fire incidence.

FIGURE 42. Successional stages of Atlantic white cedar bogs in South Carolina in relation to fire and moisture conditions (*adapted from Garren, 1943*).

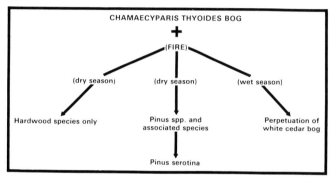

The incidence and severity of the fire factor is of great importance in successional stages of sandhill communities, especially in bogs dominated by Atlantic white cedar. This tree germinates and grows best after fires have removed cover vegetation, because young seedlings are not able to survive under dense cover, especially a cover of hardwoods (Harlow, 1968). Garren (1943) reports that if fire occurs during the wet season after mature seeds have been embedded in the peat, nearly pure stands of Atlantic white cedar will be produced. However, if fires occur during drier periods, the Atlantic white cedar-dominated bog may not be maintained.

If protected from fire, the Atlantic white cedar forest will give way in the natural course of succession to a bog community, dominated by hardwoods or semi-evergreen species (Buell and Cain, 1943) such as red bay and sweet bay.

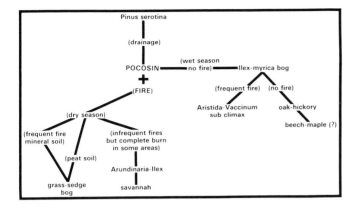

FIGURE 43. Successional stages of pocosins in South Carolina in relation to fire and moisture conditions (*adapted from Garren, 1943*).

Natural Vegetation System

Turkey Oak Barrens

For a person not interested in botany, a walk through the most xeric stages of the fall line sandhills would probably be very boring as well as very hot during the summer. He would find an open canopy of tall longleaf pines (*Pinus palustris*) overtopping numerous turkey oaks (*Quercus laevis*), as well as a few Margaret's oaks or scrubby post oaks (*Q. margaretta*), bluejack oaks (*Q. incana*), blackgums (*Nyssa sylvatica*), and persimmon (*Diospyros virginiana*). Normally very little else would be found in the arborescent stratum, although sandhill black cherry (*Prunus serotina* var. *alabamensis*) occurs in a few localities.

Shrubby ericaceous species might be sparkleberry (*Vaccinium arboreum*) and deerberry (*V. stamineum*), with rosemary (*Ceratiola ericoides*), which is relatively localized, present also. St. Andrews cross (*Hypericum hypericoides*), sand myrtle (*Leiophyllum buxifolium*), and a legume called bristly locust (*Robinia hispida*) probably would be seen, and in some locations poison oak (*Rhus toxicodendron*) and dwarf

or winged sumac (*R. copallina*). A shrubby form of goldenrod (*Solidago pauciflosculosa*) might also be found to be locally abundant.

Herbaceous plants would be even scarcer. Characteristic species might be wire plant (*Stipulcida setacea*), tread softly (*Cnidoscolus stimulosus*), sticky foxglove (*Aureolaria pectinata*), gerardia (*Agalinis setacea*), Carolina ipecac (*Euphorbia ipecacuanhae*), several species of wiregrass (*Aristida* spp.), and the somewhat shrubby jointweed (*Polygonella polygama*). Common also would be prickly pear (*Opuntia compressa*) and a few yuccas or Spanish bayonets (*Yucca filamentosa*). The last two species are not really herbaceous, yet they are not shrubs either. Ferns and other cryptogams present include bracken fern (*Pteridium aquilinum*), sand spikemoss (*Selaginella arenicola*), and several small lichens, the most predominant being reindeer moss and British soldier, both of the genus *Cladonia*. In a very few locations one might even find a lichen called old man's beard (*Usnea* spp.) clinging to dead shrubs.

It is this relative sparsity of vegetation and the large expanses of bare sand that allow many people to refer to these regions as turkey oak barrens. However, on closer inspection, one finds that these areas are not "barren" from a floristic and phenological standpoint, since the area is quite beautiful during the peak blooming months of July through September and very striking in late autumn because of the red coloration of the turkey oaks against the tall, green longleaf pines.

The flora of the most xeric sandhill areas probably is determined by complex interactions of numerous factors. Soils are low in moisture holding capacity, low in fertility and organic matter, and usually moderately to strongly acid. Constant leaching of the soil as well as the uptake of nearly all available nitrogen results in an extremely high C/N (carbon/

FIGURE 44. An east-facing bluff along the fall line. Dominant trees are longleaf pine (*Pinus palustris*), with an occasional sweetgum (*Liquidambar styraciflua*) and blackgum (*Nyssa sylvatica*). The most interesting aspect is the shrub layer, which is a dense layer of sand myrtle (*Leiophyllum buxifolium*), along with mountain laurel (*Kalmia latifolia*), inkberry (*Ilex glabra*), and ti-ti (*Cyrilla racemiflora*).

FIGURE 45. A scrub oak barren with an open canopy of longleaf pines and many scrubby turkey oaks (*Quercus laevis*). Ground cover is lacking except for numerous patches of fruticose lichens. The few herbaceous plants are sticky foxglove (*Aureolaria pectinata*), wiregrass (*Aristida* spp.), and broomsedge (*Andropogon* spp.).

nitrogen) ratio, low total nitrogen content, low bacterial count, and as a result of all this, low CO_2 evolution. This high C/N ratio has also been found to stimulate woodiness of roots and to cause an increase in the root/shoot ratio (Duke, 1961). The above mentioned factors, along with the inability of most plants to grow rapidly enough through the sand to reach available moisture, result in a flora that is relatively constant and unique.

By some adaptive means, morphologic, physiologic, or both, these xerophytic plants have the ability to withstand the recurrent sandhill drought and periodic fires. Although the means by which these plants accomplish this purpose are manifold, there are several methods that warrant discussion. These methods include groups such as tortifoliates (plants that reorient their leaves), ephemerals, succulents, and plants whose leaf orientation and morphology result in a decrease in transpiration.

FIGURE 46. A small turkey oak (*Quercus laevis*) seedling. The vertical orientation of the leaves is an adaptation that reduces light and heat absorption from high insolation and from high reflection from the white sand.

Turkey oaks are the most striking of the tortifoliates. Petioles of these plants orient their leaves in a vertical manner especially in young plants; therefore, they are able to reduce the amount of direct sunlight absorbed and also to reduce the impact of light and heat from soil reflection at the time of day when light is most intense. In addition, turkey oak acorns contain enough moisture for germination and rapid elongation of the taproot, thereby enabling the seedling to survive during its first year of growth. Other tortifoliates include false indigo (*Baptisia cinerea*), milkweed (*Asclepias humistrata*), and two goldenrods (*Solidago odora* and *S. tortifolia*).

Two other groups are able to reduce transpiration by either having plicate leaves [as in thoroughwort (*Eupatorium album, E. leucolepis*), verbena (*Verbena carnea*), and many legumes and grasses] or by having a succulent habit [as in the milkweeds (*Asclepias* spp.) and prickly pear cactus]. In addition, many sandhill plants have narrower leaves than closely related non-sandhill species. Examples of this are wild lettuce (*Lactuca graminifolia*), beard grass (*Gymnopogon brevifolius*), ironweed (*Vernonia angustifolia*), and spurge (*Euphorbia gracilior*). Furthermore, a number of other species, such as turkey oak and members of the genus *Vaccinium*, are capable of sprouting from underground shoots after a fire, thereby increasing the shrubbiness of the flora.

Longleaf pines are also well adapted for reproduction in the sandhills because of the copious amount of endosperm contained in the seed. The energy thus stored is enough to enable the young plant to grow a deep taproot the first year. In addition, longleaf pines seem to have broader ecological tolerances than other species such as turkey oak, allowing the pines to succeed in more diversified habitats (Wells and Shunk, 1931).

Wells and Shunk (1931) also reported a number of xerophytic adaptations of herbaceous and suffrutescent plants. They found that the Carolina ipecac (*Euphorbia ipecacuanhae*) develops a taproot very early in the year; later, the taproot becomes a water storage organ. Jointweed (*Polygonella polygama*) has a woody base and drops its leaves at flowering time, thereby conserving moisture. Spikemoss (*Selaginella arenicola*) has a few fine primary roots that begin branching profusely roughly one centimeter (0.4 in.) below the surface. On barren areas of sand in direct sunlight, *Selaginella* may be found around clumps of reindeer moss (*Cladonia sylvatica*). This arrangement was not only found to keep the soil moisture level slightly above that of the surrounding sand, but also to exert a marked effect in almost totally absorbing the water of light rains and dew. Wire plant (*Stipulicida setacea*) carries out its vegetative growth in early spring when moisture conditions are more favorable. On less xeric sites wire-grass may be found to have a high degree of

FIGURE 47. An excellent growth of a fruticose lichen (*Cladonia*) and sand spikemoss (*Selaginella rupestris*) in a turkey oak-dominated sandhill community. The spikemoss tends to form ring-shaped growths.

FIGURE 48. A small colony of sand spikemoss (*Selaginella rupestris*) surrounded by fruticose lichens (*Cladonia*). These two species are the chief colonizers of expanses of dry sand in the sandhills. As organic matter builds up, herbaceous plants and grasses begin to appear. The leaves indicate that the principal woody species are longleaf pine (*Pinus palustris*), turkey oak (*Quercus laevis*), and bluejack oak (*Q. incana*).

cellulose and lignin in nearly all non-photosynthetic tissue, whereas dwarf huckleberry (*Gaylussacia dumosa*) and milk pea (*Galactia volubilis*) have an extremely thick cutin layer on their upper epidermis and also a deep, water-holding taproot system. A locust (*Robinia nana*), which also extends into the lower piedmont, has highly functional leaf pulvini that cause leaflets to lie face-to-face during the daylight hours, thus reducing transpiration.

Not all the sandhill areas are as barren as the one just described. In many areas the soil is found to have a grayish-brown, loamy, sand surface layer with much more organic matter.

Herbaceous plants other than the previously mentioned species occur at regular intervals. Those most prevalent are queen's delight (*Stillingia sylvatica*) of the Euphorbiaceae and several legumes such as lupine (*Lupinus villosus*), milk pea (*Galactia volubilis*), and sensitive brier (*Shrankia microphylla*). The latter two are not restricted to turkey oak barrens but are present in many communities.

Sandhill milkweed (*Asclepias humistrata*) is scattered, and small clumps of spiderwort (*Tradescantia rosea* var. *graminea* and *T. rosea* var. *rosea*) are found in the shade near the base of scrub oaks. *Tradescantia rosea* var. *graminea* is more characteristically found in the fall line sandhills, whereas var. *rosea* is generally restricted to the lower piedmont and sandy areas of the southeastern corner of the state.

Probably the species of the Asteraceae, more than any other family, characterize the herbaceous flora of the sandhills. One generally finds the greatest blooming activity of this group in late summer and very early autumn, thus coinciding with the blooming of the Leguminosae, the sandhill's second most diverse family. Typical composites are thistle (*Carduus repandus*), dogfennel (*Eupatorium compositifolium*), golden aster (*Heterotheca gossypina*, *H. mariana* and *H.*

FIGURE 49. Typical sandhill vegetation. Turkey oak (*Quercus laevis*) and bluejack oak (*Q. incana*) are the principal woody species. The small shrub in the left foreground is rosemary (*Ceratiola ericoides*), a shrub localized in the sandhills. Ground cover consists of fruticose lichens, broomsedge (*Andropogon* spp.), and wiregrass (*Aristida* spp.).

FIGURE 50. A scrub oak barren of the sandhills. This area is dominated by longleaf pine (*Pinus palustris*), turkey oak (*Quercus laevis*), bluejack oak (*Q. incana*), and blackjack oak (*Q. marilandica*). Although this area is not as xeric as turkey oak barrens, it is still dry enough to support a few well-developed rosemary (*Ceratiola ericoides*) behind the longleaf pines (right). Ground cover is still relatively sparse, but in most cases it is much better developed than in turkey oak barrens.

graminfolia), and blazing star (*Liatris tenuifolia* and *L. secunda*).

Along roadsides and disturbed places are found sticky foxglove (*Aureolaria pectinata*) and purple gerardia (*Agalinis purpurea*), poor-Joe (*Diodia teres*) and cottonweed (*Froelichia floridana* and *F. gracilis*). Spanish moss (*Tillandsia usneoides*) is present in scattered localities but is usually not abundant. The only fern present in abundance is bracken fern (*Pteridium aquilinum*).

Scrub Oak Barrens

In areas of the sandhills where moisture conditions are slightly more favorable or where a clay subsoil is present (Wells and Shunk, 1931), a characteristic change in vegetation occurs. The dominant gymnosperm over most of the region is still longleaf pine (*Pinus palustris*), but slash pine (*P. elliottii*) is very predominant in the southeastern corner of the state. In all areas though, the oak species have changed dramatically. No longer is turkey oak the most prevalent hardwood. Here there is an increase in other scrub oaks such as bluejack (*Q. incana*), blackjack (*Q. marilandica*) and Margaret's oak (*Q. margaretta*). In areas even more favorable, southern red oak (*Q. falcata*) gains in dominance, and it becomes increasingly more important in the red hills and lower piedmont. Other tree species present include persimmon (*Diospyros virginiana*), sassafras (*Sassafras albidum*), and blackgum (*Nyssa sylvatica*).

Shrubby species are different also from those of the turkey oak barren. Characteristically they include holly (*Ilex opaca*) and in a few cases yaupon holly (*I. vomitoria*), wax myrtle (*Myrica cerifera*), sparkleberry (*Vaccinium arboreum*), deerberry (*V. stamineum*), huckleberry (*Gaylussacia frondosa*), dwarf huckleberry (*G. dumosa*), and a few others such as hawthorn (*Crataegus uniflora* and *C. flava*).

The herbaceous stratum is more pronounced in this community than in the more xeric turkey oak barrens. Within this stratum the composites still predominate, with the proportion of legumes also increased greatly over those of the more xeric areas. Most species found in turkey oak barrens are also found in scrub oak barrens, but usually have a higher frequency in the latter.

Typical composites include thistle (*Carduus repandus*), dogfennels (*Eupatorium compositifolium* and *E. capillifolium*), and blazing star (*Liatris tenuifolia*). As mentioned earlier, the Leguminosae have increased in numbers dramatically, but still have several familiar turkey oak barren species such as lupine (*Lupinus villosus* and *L. perennis*) and false indigo (*Baptisia tinctoria*). Yellow flowering legumes include partridge pea (*Cassia nictitans*) and several species of rattlebox (mainly *Crotalaria purshii* and *C. angulata*). Other legumes normally seen are pencil flower (*Stylosanthes biflora*), two species of rhynchosia (*Rhyncosia reniformis* and *R. difformis*), Samson snakeroot (*Psoralea psoralioides*), many species of beggar's ticks or beggar's lice (*Desmodium* spp.), and several species of lespedeza, those most notable being *Lespedeza procumbens, L. cuneata, L. repens*, and bush clover (*L. bicolor*). Goat's rue (*Tephrosia virginica*) is often found in large clumps along with milk pea (*Galactia volubilis*) and grasses such as broomsedge (*Andropogon scoparius*) and several species of panic grass (*Panicum amarulum, P. anceps, P. laxiflorum*, and others). Locally abundant plants are St. Andrews cross (*Hypericum hypericoides*) and pine-weed (*Lechea legettii* and *L. patula*, the former usually occupying more mesic sites).

Xeric Pine-Mixed Hardwoods

In some areas of the sandhills, moisture conditions become more mesic, and the soil often develops a characteristic precipitated subsurface horizon caused by an accumulation of silicate clays. This change in soil is accompanied by a corresponding change in vegetation. It might be pointed out that most of these areas, where suitable, have seen cultivation, and that severely eroded abandoned fields do not support well-developed vegetation.

Loblolly pine (*Pinus taeda*) is usually the dominant gymnosperm, replacing longleaf pine in nearly all cases. This is especially true southwestward toward Georgia and westward into the lower piedmont. Red cedar (*Juniperus virginiana*) also becomes more evident in areas where soil is circumneutral to alkaline. Although turkey oak is still present in drier areas, the dominant oaks now are post oak (*Quercus stellata*), southern red oak (*Q. falcata*), and scrubby post oak or Margaret's oak (*Q. margaretta*). There is an increasing occurrence of other oaks as well, the most notable of these being black oak (*Q. velutina*) and water oak (*Q. nigra*) in more mesic habitats, and bluejack oak (*Q. incana*) on the more xeric sites. Hickories include pignut (*Carya glabra*) and mockernut (*C. tomentosa*). Pale hickory (*C. pallida*) is also present in some scattered localities, but is usually more prevalent on dry rocky sites of the piedmont.

Characteristic understory species include flowering dogwood (*Cornus florida*), sassafras (*Sassafras albidum*), blackgum (*Nyssa sylvatica*), and persimmon (*Diospyros virginiana*). Shrubs are dominated by sparkleberry (*Vaccinium arboreum*), deerberry (*V. stamineum*) and huckleberry (*Gaylussacia frondosa*) in most areas, and by beauty bush (*Callicarpa americana*), wax myrtle (*Myrica cerifera*), and New Jersey tea (*Ceanothus americanus*) in others. Winged sumac (*Rhus copallina*) is usually scattered throughout, especially in areas where sunlight is plentiful. The most typical vines are poison ivy (*Rhus radicans*), yellow jessamine (*Gelsemium sempervirens*), muscadine (*Vitis rotundifolia*), and greenbrier (*Smilax glauca*).

Since the canopy tends to vary from a closed to semi-closed condition, the herbaceous layer is often spotty, tending to vary with the conditions of moisture and sunlight. Grasses such as wire-grass (*Aristida* spp.), panic grass (*Panicum* spp.), and broomstraw (*Andropogon* spp.) are usually present in more xeric areas, along with several small legumes such as leadplant (*Amorpha herbacea*), false indigo (*Baptisia tinctoria*), pencilflower (*Stylosanthes biflora*) and zornia (*Zornia bracteata*). Goat's rue (*Tephrosia virginica*) is locally prevalent with another species, *Tephrosia spicata*, appearing with it in clearings and along roadsides. In many places indigo (*Indigofera caroliniana*) may be found in local abundance. Milk pea (*Galactia regularis* and *G. volubilis*) is likewise present in these areas.

In more mesic areas under trees, spotted wintergreen (*Chimaphila maculata*) may be found associated with partridge berry (*Mitchella repens*). Typical ferns include ever-present bracken fern (*Pteridium aquilinum*) and resurrection fern (*Polypodium polypodioides*). Resurrection fern is not so widespread as bracken fern, and is found as an epiphyte on limbs of large oaks.

Pocosins, Bay Forests, and Hillside Bogs

In many areas of the sandhills the fluctuating water table intersects the surface, resulting in standing water if drainage is poor. Such areas, known as pocosins, are characterized by peaty soils and dense shrub and vine growth. Pocosin vegetation may also form on mineral soil (in which case the water is always stagnant) and on floating or anchored mats of vegetation in shallow waters of bay lakes of the coastal plain.

Ground cover is principally peat moss (*Sphagnum* spp.), but it may be absent depending upon water depth. Bladderwort (*Utricularia subulata, U. inflata, U. fibrosa* and *U. biflora*), sedges (*Carex vulpinoidea, C. howei, C. lupulina,* and *C.*

intumescens), rushes (*Juncus coriaceus* and *J.biflorus*) and spike rushes (*Eleocharis obtusa, E. `tuberculosa*, and *J. biflorus*) and spike rushes (*Eleocharis obtusa, E. tuberculosa,* and *E. microcarpa*) occur in shallow areas; whereas water lily (*Nymphaea odorata*) and water shield (*Brasenia schreberi*) grow in deeper spots where *Sphagnum* is absent. Typical ferns include cinnamon fern (*Osmunda cinnamomea*), netted chain fern (*Woodwardia areolata*) and sensitive fern (*Onoclea sensibilis*).

Shrubs present are zenobia (*Zenobia pulverulenta*), fetterbush (*Lyonia lucida*), sweet pepperbush (*Clethra alnifolia*), and hollies (*Ilex glabra* and *I. coriacea*). Sheepkill (*Kalmia angustifolia*) is present in a few localities in the northern part of the province, but it is much more prevalent north of the state. Virginia willow (*Itea virginica*) is often found around the edge, and it is associated with cane (*Arundinaria gigantea*), fetterbush (*Leucothoe racemosa* and *L. axillaris*), myrtle (*Myrica cerifera*), highbush blueberry (*Vaccinium corymbosum*), creeping blueberry (*V. crassifolium*), and swamp azalea (*Rhododendron viscosum*).

The dominant tree is pond pine (*Pinus serotina*), which occurs with a high frequency of red maple (*Acer rubrum*). Other arborescent species are red bay (*Persea borbonia*), sweet bay (*Magnolia virginiana*), and some loblolly pine (*P. taeda*). Chief lianas are greenbrier (*Smilax laurifolia, S. glauca, and S. walteri*), muscadine (*Vitis rotundifolia*) and summer grape (*V. aestivalis*).

Closely resembling pocosins are bay forests, which have a relatively open canopy with a shrub layer similar to the pocosin. Several types of bay forests exist, depending upon conditions of the soil and the arborescent species present. In areas subjected to periodic fires while the soil is still submerged, stands of Atlantic white cedar (*Chamaecyparis thyoides*) may develop. Associated tree species are tulippoplar (*Liriodendron tulipifera*), red maple (*Acer rubrum*), and

pond pine (*Pinus serotina*). The understory is usually composed of saplings of gum (*Nyssa sylvatica* var. *biflora*), sweet bay (*Magnolia virginiana*), red bay (*Persea borbonia*), red maple, and leatherwood or ti-ti (*Cyrilla racemiflora*). Shrub layer composition in bays closely resembles that of pocosins, although it may not be as dense. Herbaceous species include peat (*Sphagnum* spp.), partridge berry (*Mitchella repens*), and wild ginger (*Hexastylis arifolia*). The last two species usually appear in drier areas. Greenbrier (*Smilax* spp.) and cane (*Arundinaria gigantea*) are also common locally. In addition, a rare blueberry (*Vaccinium sempervirens*) possibly has its entire distribution along the headwaters of Scouter Creek in Lexington County (Rayner, 1978).

A second type of bay forest exists where fires are infrequent. This type is dominated by red bay (*Persea borbonia*), gum (*Nyssa sylvatica* var. *biflora*), sweet bay (*Magnolia virginiana*), and bald cypress (*Taxodium distichum*). Loblolly bay (*Gordonia lasianthus*) often appears as an ecotonal species, although at other times it may be a dominant or codominant.

Sandhills often contain areas with sloping topography and clay hardpan. Because of these conditions and the accompanying seepage, the black humic or peaty soil is nearly always wet. These areas are more protected from fire than pocosins and may form an ecotone between pocosins and bay forests and the more mesic sandhills. Common alder (*Alnus serrulata*) and poison sumac (*Rhus vernix*) are probably the only species in the area not found in either extreme, and thus may be considered as characteristic species of hillside bogs.

FIGURE 51. A well-developed bay forest in Lexington County. The canopy is composed of Atlantic white cedar (*Chamaecyparis thyoides*), loblolly bay (*Gordonia lasianthus*), tulip-poplar (*Liriodendron tulipifera*), and pond pine (*Pinus serotina*), whereas the understory is dominated by saplings of red maple (*Acer rubrum*), tulip-poplar, gum (*Nyssa sylvatica*), sweet bay (*Magnolia virginiana*), and red bay (*Persea borbonia*). The shrub layer is an almost impenetrable mass of ti-ti (*Cyrilla racemiflora*), fetterbush (*Lyonia lucida*), hollies, and alders intertwined with various vines, including cross-vine (*Anisostichus capreolata*), climbing hydrangea (*Decumaria barbara*), and greenbrier (*Smilax* spp.).

FIGURE 52. An Atlantic white cedar bog in Lexington County. This is one of the few remaining Atlantic white cedar (*Chamaecyparis thyoides*) bogs in the State. This species is characteristic of freshwater swamps and bogs that occur on peat-covered soils underlain by sand. The border also contains tulip-poplar (*Liriodendron tulipifera*) and loblolly bay (*Gordonia lasianthus*), with ti-ti (*Cyrilla racemiflora*), red maple (*Acer rubrum*), and red bay (*Persea borbonia*) along the inner border. Herbaceous species include various species of pitcher plants (*Sarracenia rubra* and *S. purpurea*), sundew (*Drosera intermedia*), bur-reed (*Sparganium americanum*), and arrowhead (*Sagittaria* spp.). ⟶

PART FIVE: Coastal Plain Province

Physiographic Features
>Rolling Hills and Red Hills
>Limestone Sinks
>Savannahs
>Disjunct Sandhills
>Carolina Bays
>River Swamps and Ridges
>Marshes
>Maritime Strand
>Barrier Islands

Natural Vegetation System
>Turkey Oak Barrens
>Scrub Oak Barrens
>Xeric Mixed Hardwoods and Pine
>Mesic Mixed Hardwoods and Pine
>River Bluffs and Beech Ravines
>Magnolia Forest
>Florida Scrub
>Bald Cypress-Tupelo Swamps
>Upland Swamps
>Major River Bottoms
>Oxbow Lakes
>Stream Banks
>Hardwood Bottoms
>Hardwood Cane Bottoms

>Pond Cypress Savannahs
>Pine-Toothache Grass Savannah
>Pine-Wiregrass Savannah
>Oak Savannah
>Lotic Water Communities
>Limestone Sinks
>Old Growth Pine Communities
>Coastal Dunes
>Dune-Shrub Communities
>Maritime Forest
>Salt Marshes
>Salt Marsh Border Zonation
>Brackish Marshes
>Freshwater Marshes

Physiographic Features

SOUTH CAROLINA'S COASTAL PLAIN PROVINCE from the Sandhill Province eastward is a flat to gently dissected area dominated to a large extent by gymnosperms of one kind or another. That this form of plant life predominates is not unexpected, since the area as a whole belongs to the southeastern coniferous forest association. In most areas hardwoods rarely attain complete dominance because of the high incidence of fire. Exceptions to this, however, are well represented by maritime forests and wide-spread hardwood river bottoms along major rivers such as the Congaree, Lynches, the Pee Dee, and the Savannah. Exceptions may also be noted along the inner border often known as the red hills, where the coastal plain intergrades with the Sandhill and Piedmont Provinces.

In general however, the coastal plain may be arbitrarily divided into three somewhat characteristic regions: outer coastal plain and maritime strand, middle coastal plain, and inner coastal plain.

Outer Coastal Plain and Maritime Strand

South Carolina is blessed with many miles of beautiful public beaches. In addition, the coast has many square miles of dunes, salt marshes, maritime forests, and barrier islands. Many areas are largely undisturbed, but others are succumbing to commercial and residential development and pollution.

FIGURE 53.   A very picturesque maritime forest scene on Capers Island. Live oaks (*Quercus virginiana*) and palmetto trees (*Sabal palmetto*) often dominate these areas (*courtesy S.C.W.M.R. Department.*)

FIGURE 54. A grove of palmetto trees (*Sabal palmetto*) along a canal on Fripp Island, Beaufort County. Many such groves exist along the maritime section of the southeastern coastal counties. The small root system subjects these plants to easy windthrow during the frequent storms along the coast.

Vegetation in these areas is drastically influenced by salt concentration and wind and salt spray. It seems to most investigators that the outer areas of the coastal plain are distinctive enough to warrant separate classification from the remainder, and therefore, we shall refer to them as the maritime strand.

In this maritime strand, it also seems only natural to refer to the included communities as salt spray communities, since the growth of vegetation is directly affected by wind-carried salt spray and/or salt concentration in marsh habitats.

Whereas dune communities are directly affected by salt spray and wind, estuaries are strongly affected by tidal action and freshwater drainage from rivers and land. In many ways they may even be considered as an ecotone between the freshwater and marine habitats, although many of their biological and physical attributes are not transitional but, in their own way, unique. Because of their uniqueness, estuaries have drawn much interest from researchers and conservationists, and classification schemes for these unique habitats are numerous. Most current classifications deal with one or more of the following criteria: geomorphology, water circulation and stratification, and systems energetics.

In examining naturally formed South Carolina estuaries, it seems that most were at one time or another formed from a basin enclosed by a chain of off-shore bars or barrier islands. This barrier is not continuous; it is broken by inlets in many places, thus assuring the marshes a free connection with the Atlantic Ocean. In fact, the near absence of barrier islands north of the Pee Dee River, largely in Horry County, is thought to be a result of progressive effects of erosion (Cooke, 1936).

Middle Coastal Plain

The remainder of the Coastal Plain Province can be roughly divided into two zones in respect to topography. The middle-coastal plain is relatively flat, whereas the inner coastal plain exhibits a relief generally characterized by river swamps and wide ridges. These low ridges become more pronounced as one approaches the Sandhill and Piedmont Provinces; and in areas south of the Santee River, a different soil profile results in a reddish hue, adding to the connotation of red hills.

Incorporated into the relatively flat lands of the mid-coastal plain are freshwater marshes, savannahs, river

FIGURE 55. A northward aerial view of Turtle Island, Jasper County, a barrier island characteristic of many along the coast. It has approximately 1590 acres of low marsh dominated by cordgrass (*Spartina alterniflora*) and by bulrushes (*Scirpus robustus*) in less saline areas. There are 150 acres of wooded upland and beach area along the eastern and southeastern borders. Marshland separates the island from Daufuskie Island on the horizon.

swamps and ridges, disjunct sandhills, limestone sinks, and of course, mesic woodlands and the peculiar Carolina Bays.

Our shallow freshwater marshes are usually dominated by cattails, grasses, sedges, and rushes, and in deeper areas algae such as *Spirogyra, Hydrodicton*, and *Nostoc* may be abundant. Thick mats of peat often form a substrate for herbs and small shrubs.

River swamps on the other hand are wooded wetlands that seem to exhibit a successional step from marshlands to mesic forests. Our swamps are relatively easily divided into two types: deep water and shallow water. Deep water swamps occur most extensively along the flood plains of larger rivers where they are characterized by cypress and tupelo-gum, and a relative sparsity of herbaceous vegetation. However, these deep water swamps do support an abundance of epiphytes such as Spanish moss and mistletoe. Shallow water swamp vegetation ranges from shrubby willows, alders, and button bush to oaks and maples, with overcup oak and water hickory as commonly associated species.

FIGURE 56 (opposite page). A well-developed floating mat vegetation system. This old mill-pond has developed a very thick layer of peat on which herbaceous and shrubby vegetation grows with considerable abundance. In addition to a small channel through the center of the mat, there is still a layer of water under much of the vegetation.

FIGURE 57. Many low flatland bogs along the coast show zonation in respect to substrate solidity, substrate composition, and elevation above the water table. This site in Georgetown County demonstrates variations from standing water dominated by water lilies (*Nymphaea odorata*) and maidencane (*Panicum hemitomon*) to small elevations dominated by shrubs and pines.

Mesic woodlands and disjunct sandhills offer much the same vegetation pattern as do their counterparts in other sections of the region. Moisture and soil conditions and elevation above the water table account for their occurrence.

Lime sinks also dot the coastal plain in areas of large deposits of calcareous sediments. These depressions may be quite small or may range upwards of several acres. Depths also may vary with the amount of dissolution of substrate and the amount of vegetational succession. A few areas such as some near Santee State Park have worn away enough limestone to produce large holes and deep caves. In all cases, however, vegetation is typified by calciphytes, and depending upon the size of the sink, zonation may not be evident.

Finally, but surely not least in interest, are the so-called Carolina Bays. These curious depressions dot the coastal plain of both Carolinas, and they all have a similar shape and orientation. Porcher (1966) denotes three general physiographic features of the bays that he studied: a general ovate to elliptical shape; a general northwest to southeast directionalism (in his study most had a long axis running from SE 145° to NW 325°); and a sand ridge or ridges in the southeastern quadrant.

Depth of water varies from small, shallow lakes supporting very little anchored vegetation to shallow areas with a preponderance of peat, herbaceous species, and small shrubs. In nearly all cases, typical pocosin species predominate, with the outer edges having a border of typical bay forest species. It is likely that location and the presence of these bay forests weighed heavily in the adoption of the somewhat parochial name of Carolina Bays.

Origin of these elliptic depressions has also drawn considerable interest. Many ideas have been presented incorporating substrate dissolution and other weathering processes; however, one theory was proposed by Melton (1950). He postulated that it was possible that a meteorite shower impact in this area occurred at any one of several times. An outline story of Carolina Bay origin and history involving their catastrophic origin was presented by Wells and Boyce (1953). After the initial formation of a bay, they postulated that the bay rapidly filled in with silt from the surrounding devastated region. Subsequent development of the aquatic peat community to a point where a swamp forest of some kind was established was followed in post-glacial time by extensive reduction of the forests to shrub bogs through local deep burning of the peat. In addition, a number of bays near deep river drainages formed small lakes within their borders.

FIGURE 58. An aerial view of Woods Bay State Park, one of the largest of the Carolina Bays that dot the coastal plain. They all seem to be oriented in a southeast-northwest direction with one or more sand ridges in the southeastern quadrant. The depth of water varies from small, shallow lakes supporting very little anchored vegetation to shallow areas with a preponderance of peat, herbaceous species, and small shrubs (*courtesy S.C.W.M.R. Department*).

FIGURE 59. Woods Bay vegetation. Throughout this and other Carolina Bays, there is, depending upon water depth, a growth of bald
cypress (*Taxodium distichum*), tupelo gum
(*Nyssa aquatica*), and other swamp hardwoods.
Ground cover is a mixture of grasses, sedges,
and other herbaceous plants with floating anchored aquatics predominating in areas of standing water (*courtesy S.C.W.M.R. Department*).

Cooke (1936), on the other hand, suggests that the bays were formed by wave action on old sandy shores of abandoned inland waterways. This wave action naturally occurred before the bodies of water had been converted into swamps, and he also points out that their elliptical shape and long axes were generally at right angles to the coastline as if winds blowing from the ocean has set in motion and directed the waves that shaped them. The directional data reported by Porcher also tend to support this hypothesis.

However, Carolina Bays are not the only bays present in the area. Johnson (1942) found various irregular bays or craters not considered to be Carolina Bays, although they show some of the same physiographic and vegetational features.

Inner Coastal Plain

Although the inner coastal plain is very similar in many aspects to the middle-coastal plain, relief is quite different. The latter has a surface that is basically one of primary topography, whereas inner coastal plain topography is one showing considerable weathering—a fact well demonstrated by the many steep bluffs along major rivers. Relic surfaces, which regionally reflect alluvial fan or deltaic shaped landforms, can be easily visualized in examining the topography (Colquhoun, 1969).

Mesic woodlands occur in greater quantities because of the increase in uplands; however, freshwater marshes and bay forests still occur, and often have a sharper transitional structure to surrounding areas.

On the rolling hills just seaward of the sandhills it may be generally noted that there is a more mesic flora than that of the Sandhill Province, yet distinctively more xeric than the mid-coastal plain. In most areas this can be traced to the fact that these areas are considerably above the water table, and

FIGURE 60. The mosquito fern (*Azolla caroliniana*), shown here in a small farm pond, is just one of many plants that have migrated into the State and have reproduced in such quantity as to become a nuisance.

FIGURE 61. An extremely dry white sand disjunct sandhill in Allendale County completely dominated by turkey oak (*Quercus leavis*) and other sandhill species.

that the soil is more porous, less loamy, and lower in organic matter content than flatter portions of the coastal plain.

Vegetational response to this myriad of edaphic conditions has led to a remarkable mosaic of plant communities. The coastal plain is generally considered by most people to be a monotony of southeastern evergreens, but it must be realized that fluctuations in soil structure, nutrient content, and water content have produced a composit flora much more varied than either the Mountain or Piedmont Provinces.

Natural Vegetation System

Turkey Oak and Scrub Oak Barrens

Discussion of turkey oak and scrub oak barrens will be limited here in that these disjunct sandhill communities are vegetatively similar to communities of the Sandhill Province. However, one point which should be noted is that disjunct sandhills may be found in many unexpected locations. Any place where coarse sand is high enough above the water table may, over a period of time, develop typical sandhill vegetation.

Two site descriptions will serve to elucidate this point. In lower Orangeburg County there is a small community known as Sandy Island. Most of the surrounding area is swampland, but there are large expanses of disjunct sandhill communities scattered throughout. One rather large site has, in addition to typical sandhill vegetation, a rather large population of rosemary (*Ceratiola ericoides*) and a lichen called old man's beard (*Usnea* spp.). Both species are most uncommon in this section of the state. Ground cover is dominated by very extensive clumps of reindeer moss (*Cladonia* spp.).

A second site is located in Allendale County in an area known appropriately as "white sand ridge." This ridge is approximately a half mile (802.6 m) long and 150 yards (136.8 m) wide, and is composed of unusually white sand extremely low in nutrients and moisture. Vegetation here is sparse, but what there is again typifies the Sandhill Province.

Xeric Mixed Hardwoods and Pine

It can generally be noted that the red hills support more mesic flora patterns than do the sandhills, with much of this being attributed to the heavier loamy soils having a greater

FIGURE 62. A disjunct sandhill habitat in Horry County dominated by scrubby live oaks (*Quercus virginiana*) and pine with a very sparse shrub and herbaceous layer. Unlike most sandhill areas, humidity is high enough to allow a copious growth of Spanish moss (*Tillandsia usneoides*).

FIGURE 63. Many of the drier sections of the outer coastal plain resemble the scrub oak barrens of the Sandhill Province in species composition. These areas are dominated by longleaf pine (*Pinus palustris*) and various xeric oaks such as post oak (*Quercus stellata*), Margaret's oak (*Q. margaretta*), bluejack oak (*Q. incana*), and turkey oak (*Q. laevis*). Shrub and herbaceous layers very closely resemble those of the sandhills.

water holding capacity and a greater organic matter content. The most xeric communities of the red hills produce a flora generally similar to the most mesic areas of the sandhills, namely, xeric mixed hardwoods and pine, characterized by loblolly pine (*Pinus taeda*) as the dominant gymnosperm and post oak (*Quercus stellata*), southern red oak (*Q. falcata*) and scrubby post oak (*Q. margaretta*) as the oaks most commonly seen. Principal hickories are mockernut (*Carya tomentosa*) and pignut (*C. glabra*). Turkey oak (*Q. laevis*) is still present on most xeric sites, and red cedar (*Juniperus virginiana*) becomes more prominent where areas blend with the piedmont.

This community type is most prevalent in the better drained ridges and fine sand ridges; it is poorly developed in heavily eroded soils and pinelands. It is hard to truly characterize this community other than to describe the dominant arborescent stratum because of local soil changes that affect principally the herbaceous stratum, and to a lesser degree, the understory and shrubby species. In many areas this type succeeds worn out cotton fields. The relative infertility of this soil is probably a major factor limiting further development of hardwoods.

Mesic Mixed Hardwoods and Pine

Midslope on red hills shows a characteristically mesic change in vegetation. Once again, even with the occurrence of many subtypes, the vegetation cover may be generally classified as mesic mixed hardwoods and pine. A similar community type may also be found on fine sand ridges and well-drained flatwoods that are mostly protected from fire.

In general, loblolly pine (*Pinus taeda*) is replaced by more mesic hardwoods than in the preceding type. The dominant oak now is white oak (*Quercus alba*), either solely dominant or in combination with loblolly pine.

Associates include a mixture of xeric and mesic species. Those principally found in the overstory are sweetgum (*Liquidambar styraciflua*), beech (*Fagus grandifolia*), southern red oak (*Q. falcata*), post oak (*Q. stellata*), mockernut hickory (*Carya tomentosa*), and southern sugar maple (*Acer saccharum* spp. *floridanum*). The presence of tulip-poplar (*Liriodendron tulipifera*) generally distinguishes this community from that of hardwood bottoms.

Understory is dominated by flowering dogwood (*Cornus florida*), sourgum (*Oxydendrum arboreum*), redbud or Judas tree (*Cercis canadensis*), and a few smaller species including holly (*Ilex opaca*) and leatherwood or ti-ti (*Cyrilla racemiflora*). Herbaceous flora is generally varied, however it does include many species of the xeric woodlands as well as those more prevalent in the piedmont.

River Bluffs and Beech Ravines

Although normally limited to north-facing bluffs, where there is moist well-drained soil, beech ravines may occur in the red hills on finer sand ridges and well-developed soils of the coastal plains. It is in these coves that we find a strange mixture of piedmont and lower mountain species, as well as a few bottomland species mixed in with typical mesophytic hardwoods. Because of this unique assemblage of species the term "mixed mesophytic hardwood forest" has been applied as well as the term "beech ravine." The latter term ideally refers to the predominance of beech (*Fagus grandifolia*). Familiar local species in addition to beech are white oak (*Q. alba*), loblolly pine (*P. taeda*), mockernut hickory (*C. tomentosa*), and tulip-poplar (*Liriodendron tulipifera*).

In some instances, one finds the presence of upper piedmont species such as black oak (*Q. velutina*), red oak (*Q. rubra*), pignut hickory (*C. glabra*), and winged elm (*Ulmus alata*). Mountain laurel (*Kalmia latifolia*) is common on slopes and shrubby dogwood (*Cornus amomum*) is found in low moist places. *C. amomum* is normally limited to the Mountain and Piedmont Provinces. Umbrella tree (*Magnolia tripetala*) is another species characteristic of the piedmont and mountains that has been reported from scattered locations in the coastal plain.

Understory species include flowering dogwood (*Cornus florida*), witch hazel (*Hamamelis virginiana*), musclewood (*Carpinus caroliniana*), sassafras (*Sassafras albidum*), holly (*Ilex opaca*), redbud (*Cercis canadensis*), storax (*Styrax americana*), spicewood (*Lindera benzoin*), and strawberry bush (*Euonymous americanus*).

Vines often predominate in these areas, with the larger ones being climbing hydrangea (*Decumaria barbara*) and poison ivy (*Rhus radicans*). Although usually small in these localities, trumpet vine (*Campsis radicans*), and Virginia creeper (*Parthenocissus quinquefolia*) may be very extensive.

Herbaceous flora may show a great tendency toward upper piedmont and mountain species. Among the more beautiful flowering species are various asters, false Solomon's seal (*Smilacina racemosa*), Solomon's seal (*Polygonatum biflorum*), and goldenrod (*Solidago* spp.).

Beechdrops (*Epifagus virginiana*) are often found parasitic on the roots of beech. These herbaceous parasites have scale-like leaves without chlorophyll, and during their growing phase they have a purplish or yellow color that turns brown after flowering. Indian pipe (*Monotropa uniflora*) is another achlorophyllous flowering plant often found in these habitats. Unlike beechdrops, Indian pipe is a saprophytic member of the heath family. In addition, two other members of the heath family found rather abundantly are partridge berry (*Mitchella repens*) and spotted wintergreen (*Chimaphila maculata*).

Birthwort (*Aristolochia serpentaria*), easily distinguished

by the turpentine odor given off by freshly bruised roots, is present along with two other genera of the same family, *Asarum* and *Hexastylis*. Heartleaf or wild ginger (*Hexastylis arifolia*) is usually present in quantities in deciduous forests throughout the region, and it is also found in some areas associated with pine woods. The more pubescent wild ginger (*Asarum canadensis*) is usually restricted to rich woods. Both *Asarum* and *Hexastylis* are evergreen, have a pungent ginger aroma, and produce maroon or puce colored flowers in the leaf litter.

Crane-fly orchid (*Tipularia discolor*) is usually the only orchid present, although green adder's mouth (*Malaxis uniflora*), rattlesnake plantain (*Goodyera pubescens*), and ladies' tresses (*Spiranthes* spp.) may also be seen in scattered localities.

Among the very prevalent ferns are ebony spleenwort (*Asplenium platyneuron*), Christmas fern (*Polystichum acrostichoides*), and broad beech-fern (*Thelypteris hexagonoptera*). Rattlesnake fern (*Botrichium virginiaum*) is usually found in rich woods and thickets, but because of its small size, is usually not as noticeable as the larger species. Rattlesnake fern is unique in that it is the only one of these species that produces separate sterile and fertile fronds.

FIGURE 64. Magnolia forests, such as this one in the extreme southeastern corner of the State, are probably the northern limit of the semitropical mesophytic forest. They are characterized by magnolia (*Magnolia grandiflora*), wild olive (*Osmanthus americana*), red bay (*Persea borbonia*), laurel oak (*Quercus laurifolia*), and live oak (*Q. virginiana*). The magnolia in the extreme foreground was uprooted by erosion and tidal action in the Colleton River.

Magnolia Forest

The semitropical mesophytic forest reaches its northern limit in our southeastern counties from the vicinity of Charleston southward into Beaufort and Jasper counties. These semitropical mesophytic forests are characterized by *Magnolia grandiflora* more than any other species. Other species associated with magnolia are wild olive (*Osmanthus americanus*), red bay (*Persea borbonia*), laurel oak (*Quercus laurifolia*) and beech (*Fagus grandifolia*).

The shrub layer is normally sparse because of the closed canopy. Subcanopy species commonly found are redbud (*Cercis canadensis*), and wild olive (*Osmanthus americanus*).

Herbaceous and lianal species are quite numerous, with principal ones being those found in the surrounding regions. Spanish moss (*Tillandsia usneoides*) is very common, along with resurrection fern (*Polypodium polypodioides*) found growing on limbs of trees.

Florida Scrub

Near U.S. Route 278 east of Ridgeland and in Beaufort County there are some pine flatwoods containing slight depressions and ridges. Apparently the high water table enables several species of pines to co-exist and also enhances the development of a dense woody pocosin understory three to four feet tall in depressions. Even though fire is a frequent factor in these areas, the understory seems to maintain itself quite satisfactorily.

The overall appearance of these areas seems rather monotonous because of the open canopy of tall pines. Chief dominants in depressions are pond pine (*Pinus serotina*) and slash pine (*P. elliottii*); longleaf pine (*P. palustris*) dominates the ridges.

Understory species include myrtle leaved oak (*Quercus myrtifolia*), sweet bay magnolia (*Magnolia virginiana*), red bay (*Persea borbonia*), sassafras (*Sassafras albidum*), and wax myrtle (*Myrica cerifera*). Loblolly bay (*Gordonia lasianthus*) may also attain considerable height in pocosins.

Chief shrubs are horse sugar (*Symplocos tinctoria*), staggerbush (*Lyonia ferruginea*), inkberry (*Ilex glabra*), leucothoe (*Leucothoe axillaris*), and white alder (*Clethra alnifolia* var. *tomentosa*). Naturally the same old standbys of fetterbush (*Lyonia lucida*), winged sumac (*Rhus copallina*), hairy mountain laurel (*Kalmia hirsuta*), highbush blueberry

FIGURE 65. A Florida scrub community, in Beaufort County near Ridgeland, is dominated by a mixture of pond pine (*Pinus serotina*), longleaf pine (*P. palustris*), and slash pine (*P. elliottii*). The herbaceous layer tends to be a mixture of coastal plain flatwood and pocosin species although a few species such as the saw palmetto (*Serenoa repens*) in the foreground reach their northern limit in Beaufort and Jasper counties.

(*Vaccinium corymbosum*), and zenobia (*Zenobia pulverulenta*) are present.

While most of the herbaceous species tend to be a mixture of coastal plain flatwood and pocosin species, there are a few plants represented whose northern limit is in this area. *Vaccinium myrsinites*, *Lyonia ferruginea*, and *Serenoa repens* are likely to occur only in Jasper and Beaufort counties, while *Houstonia procumbens* is known only in three counties of the State.

Cypress-Tupelo Swamps

The familiar concept of a coastal plain swamp is one of bald cypress (*Taxodium distichum*) and water-tupelo (*Nyssa aquatica*) dominance, in an alluvial soil with open circulation of water. Normally it is flooded for most of the year. Even though both of the above species are normally indicative of flooded soil conditions, dry soil is needed for seed germination. Seedlings may, however, begin growth on floating logs and mats, then become permanently rooted at a later date. The most striking features are the nearly permanently flooded soft substratum (which is high in humus), the buttressed bases of tupelo and bald cypress, and in some areas, numerous large cypress knees. In addition to areas flooded from one to several feet deep, there may be interpersed throughout the swamp levees that support a greater array of herbaceous and shrubby species.

Cypress-tupelo swamps occur in low bottomlands along rivers, in Carolina Bays, and in deep swamps, as well as in oxbow depressions of hardwood bottoms. In each instance the dominant arboreal species, as would be expected, are bald cypress and water-tupelo. It is important from a taxonomic standpoint to distinguish *Nyssa aquatica* (water tupelo) from *N. biflora* (swamp tupelo). The latter grows principally in acidic areas where flooding is not as wide-

FIGURE 66. A typical tupelo gum (*Nyssa biflora*) swamp. Although the canopy is relatively open in late autumn, it produces a dense shade most of the year. Shade, along with standing water and general acidic conditions, precludes the establishment of many plants. However, such humid environments do afford conditions for a luxuriant growth of mosses and liverworts on the trees.

spread, in borders of cypress-tupelo swamps, and in areas with shallow stagnant water. Associates in cypress-tupelo swamps are water ash (*Fraxinus carolinianus*), red maple (*Acer rubrum*), black willow (*Salix nigra*) and water elm (*Planera aquatica*). Within the Congaree River floodplain there are small areas exhibiting nearby pure stands of water elm, which often arise after a disturbance, such as clear cutting or extensive wind damage (Rayner, 1976).

As expected, depth and period of inundation play an extremely important part in species composition. Areas where flooding is relatively constant are usually dominated solely by bald cypress and tupelo. As the depth of flooding decreases along with a corresponding decrease in the duration, more codominants such as red maple, boxelder maple (*A. negundo*), and water elm appear. Other commonly found associates in the lower coastal plain are sycamore (*Platanus occidentalis*) and cottonwood (*Populus heterophylla*).

Understory in most areas consists of red bay (*Persea borbonia*), sweet-bay magnolia (*Magnolia virginiana*), sugarberry (*Celtis laevigata*) and American elm (*Ulmus americana*). Winged elm (*U. alata*) also occurs in places, but its occurence is generally limited to upland soils along streams and rivers. Smaller species occurring frequently are papaw (*Asimina triloba*), buttonbush (*Cephalanthus occidentalis*), and American or Christmas holly (*Ilex opaca*). Two deciduous hollies are frequent also, such as winterberry (*I. verticillata*) and smooth winterberry (*I. laevigata*), as well as two evergreen hollies (*I. coriacea* and *I. glabra*) which are more commonly found in pocosins and bays. Other shrubby species found are Virginia willow (*Itea virginica*), and several ericacious individuals such as fetter-bush (*Lyonia lucida*), and male-berry (*L. ligustrina*).

Because of the frequent flooding, herbaceous species are generally few, however those that do appear are generally in abundance.

Aquatics include burweed (*Sparganium americanum*), duckweed (*Lemna* spp.), arrowhead (*Sagittaria* spp.), parrot feather or water milfoil (*Myriophyllum* spp.), and mermaid weed (*Proserpinaca* sp.). Most of these aquatics prefer an abundance of sunlight, but some species such as cattail (*Typha latifolia*) are restricted to sunny areas. Another plant that needs to be mentioned here is alligator weed (*Alternanthera philoxeroides*). Just as the American alligator is becoming more and more abundant, so is alligator weed. This small, white-flowered weed has formed extensive mats in many areas completely choking out other herbaceous vegetation and blocking channels.

Non-aquatics are more varied in number, but once again, these are usually limited to solid substrates such as levees, old stumps and logs, and sometimes floating masses of debris and vegetation. Typical herbs include lizards tail (*Saururus Cernuus*), false nettle (*Boehmeria cylindrica*), clearweed (*Pilea pumila*), various species of knotweed (most commonly *Polygonum hydropiperoides*), and greenbrier (most commonly *Smilax glauca* and *S. walteri*). Other herbs appearing in areas where sunlight is more plentiful are St. John's wort (*Hypericum gymnanthum* and *H. mutilum*) and various sedges and grasses.

Several ferns may also be located in areas where substrate permits. Most common are cinnamon fern (*Osmunda cinnamomea*), netted chain-fern (*Woodwardia areolata*), and sensitive fern (*Onoclea sensibilis*).

Probably the most common vine is poison ivy (*Rhus radicans*). Under ideal growing conditions this vine can send out shoots several inches in diameter from the main stem. A novice may even mistake it for a tree limb with trifoliolate leaves. Other lianas present are muscadine (*Vitis rotundifolia*), cross-vine (*Bignonia capreolata*), and pepper-vine (*Ampelopsis arborea*).

Of course, cypress-tupelo swamp should be without its two main epiphytes, Spanish moss (*Tillandsia usneoides*) and mistletoe (*Phoradendron flavescens*).

Typically cypress-water-tupelo swamps border into areas dominated by swamp-tupelo where associated dominants may be red maple, water hickory (*Carya aquatica*), overcup oak (*Quercus lyrata*), and swamp chestnut oak (*Q. michauxii*). There are also heavier growths of the evergreen shrubs previously listed, as well as a couple of viburnums (*Viburnum nudum* and *V. dentatum*). Herbaceous species found growing on the often peaty bottom include numerous sedges, rushes and ferns. Predominant fern species include southern lady fern (*Athyrium asplenioides*), netted chain fern (*Woodwardia areolata*), sensitive fern (*Onoclea sensibilis*), cinnamon fern (*Osmunda cinnamomea*), and royal fern (*O. regalis*).

As elevation increases above that of the tupelo swamp, better drainage is afforded and swamp-tupelo communities give way to sweetgum (*Liquidambar styraciflua*)-hardwood communities similar in composition to bottomland hardwoods.

Upland Swamps

Similarities between swamp-tupelo bogs and certain pocosins and bays of the floodplain are numerous, but in general, both have acid conditions with wet soil. Vegetation in reference to understory, shrub species, and herbaceous species is also very similar. However, upland swamps contain several different canopy species. Pond cypress (*Taxodium ascendens*) is commonly an associate along with pond pine (*Pinus serotina*). Where the substrate becomes nearly pure peat, Atlantic white cedar (*Chamaecyparis thyoides*) may be present, but as in the sandhills, its germination and early growth is dependent on fire. Slash pine (*Pinus elliottii*) also becomes more prevalent in upland swamps.

Pond cypress may attain complete dominance where moisture is variable. Excellent sites are found in Bamberg and Allendale counties as well as scattered localities throughout the southeastern portion of the state.

It may be also noted that although pond cypress is an acidophyllous plant, it is also found in areas over Santee Limestone. In these areas the soil overlying the limestone is deep enough so that relatively few basic ions are available to the shallowly rooted plants.

Major River Bottoms

Three major river systems traverse the coastal plain in South Carolina and discharge into the Atlantic: the Pee Dee, the Santee, and the Savannah. All three arise in the piedmont and contribute a large portion of their load of erosion to accretion of sedimentary deposits in shoreline and ocean environments.

During hundreds of years of meandering and slow erosion of the inter-terraces, these rivers and their tributaries have produced some of the most beautiful hardwood river bottom systems in the entire nation. Many areas such as the Congaree and Wateree River floodplains have remained in their present state for several hundred years. Moreover, certain tracts have virgin stands of timber. The meandering of these large rivers and their state of antiquity are the principal reasons why they have produced such a myriad of interesting communities.

An aerial view of wide hardwood bottoms offers a startling experience. One can see numerous oxbow lakes which were formed as the river changed direction and also numerous vegetation patterns along low areas that weave their way between masses of vegetation on slightly higher ground.

Since nearly all areas were at one time or another subject to large deposits of alluvium, moisture conditions have

FIGURE 67. Several large rivers wind their way through the flatlands of the coastal plain. These rivers are usually bordered by a wide floodplain dominated by a variety of species, depending upon the height of the water table and degree of flooding. The dominants along the Black River are a variety of hardwoods and an occasional bald cypress (*Taxodium distichum*) and tupelo (*Nyssa aquatica*). Along the margin of the river is a shrub layer of willows (*Salix nigra*), wax myrtle (*Myrica cerifera*), and hardwood saplings (*courtesy S.C.W.M.R. Department*).

played an important part in the development of the vegetation systems. There are undoubtedly many more communities than we will discuss here, but, in general they are all modifications of a few large associations. We will also approach the discussion of these systems in relation to their elevation, depth of flooding, and frequency of flooding.

FIGURE 68. An aerial view of an oxbow lake in the Congaree Swamp. Many such bodies of water are formed in the flatlands of the coastal plain when natural phenomena cause a change in a river's course. There is a large even-aged stand of hardwoods on the old river bottom, in sharp contrast to the old forest on what was once the river floodplain (*courtesy S.C.W.M.R. Department*).

Stream Banks

Stream banks and some lake extensions also provide conditions for a community different from any previously discussed. If there is frequent deposition of alluvium, thickets of willows (*Salix nigra*) and black alder (*Alnus serrulata*) may develop where canopy is not too dense. Other common species are buttonbush (*Cephalanthus occidentalis*), various species of elm (*Ulmus* spp.), and swamp cottonwood (*Populus heterophylla*).

Hardwood Bottoms

A majority of sites in large sprawling bottomlands are dominated by hardwoods in one form or another. There is obviously a great degree of variation in flooding and moisture content of these areas and, to a lesser degree, changes of elevation. In response to these variations, different species attain dominance, some in pure stands, others in mixed stands.

Because of the amount of alluvium deposited and the normally high soil moisture content, sweetgum (*Liquidambar styraciflua*) is quite often found to be dominant, or at least to share a major part of the dominance. Diameters of these trees may range up to several feet, and heights may range to more than 100 feet. Associated dominants most frequently found are green ash (*Fraxinus pennsylvanica*), American elm (*Ulmus americana*), overcup oak (*Quercus lyrata*), and laurel oak (*Q. laurifolia*). Sycamore (*Platanus occidentalis*), river birch (*Betula nigra*), and cottonwood (*Populus heterophylla*) may also gain dominance in more moist areas. Swamp Spanish oak (*Q. falcata* var. *pagodaefolia*), swamp chestnut oak (*Q. michauxii*), and swamp red oak (*Q. shumardii*) may also grow to be rather large trees.

Loblolly pine (*Pinus taeda*) is the only gymnosperm normally attaining canopy size. Since pines do not normally reproduce successfully in dense shade, individual specimens represented are usually quite old. Areas such as the Congaree Swamp contain many such trees, often as much as five feet in diameter.

Understory is dominated by hackberry or sugarberry (*Celtis laevigata*), holly (*Ilex opaca* and *I. decidua*), musclewood (*Carpinus caroliniana*), and of course swamp dogwood (*Cornus stricta*), and the ever-present hawthorne (*Crataegus* sp.). Small specimens of canopy dominants may also be

FIGURE 69. Impressive settings are common in many virgin areas of the Congaree Swamp. Loblolly pines (*Pinus taeda*) are often several hundred years old and are only present because they were established well before the onset of the dense hardwoods. Very few young pines are found under a dense canopy. In areas where a catastrophe has caused an opening in the canopy, there can be enough sunlight to allow seedling growth (*courtesy S.C.W.M.R. Department*).

FIGURE 70. A hardwood-cane bottom in the Claytor Swamp. Cane (*Arundinaria gigantea*) may become a dominant factor in alluvial hardwood bottoms where a disturbance such as fire has caused the canopy layer to be quite open. Commonly associated with cane are white oak (*Quercus alba*), beech (*Fagus grandifolia*), American holly (*Ilex opaca*), and various other hardwood bottom species.

found as well as spicebush (*Lindera benzoin*) and an occasional red mulberry (*Morus rubra*). Smaller shrubs include strawberry bush (*Euonymous americanus*) and Virginia willow (*Itea virginica*).

The forest floor in these bottoms is dominated by grasses, sedges, and small herbaceous species, as well as a few ferns. Characteristic grasses include river oats (*Uniola laxa*), and panic grass (*Panicum agrostoides*). Carex (*Carex typhina*), false nettle (*Boehmeria cylindrica*), buttonweed (*Diodia virginica*), partridge berry (*Mitchella repens*), dayflower (*Commelina virginica*), and knotweed (*Polygonum setaceum*) are also very common.

Ferns are not nearly as common as other herbaceous plants, but in most areas sensitive fern (*Onoclea sensibilis*) is most prevalent. Southern lady fern (*Athyrium asplenioides*) can also be found growing on, and subsequently sprawling over, old moss covered stumps and fallen logs.

Chief lianas would undoubtedly include poison ivy (*Rhus radicans*); in some places it literally covers the ground. Numerous other lianas are present, but wild grape (*Vitis rotundifolia*), supple-Jack (*Berchemia scandens*) climbing hydrangea (*Decumaria barbara*), and peppervine (*Ampelopsis arborea*) are the most frequent.

The two most easily seen epiphytes are Spanish moss (*Tillandsia usneoides*) and mistletoe (*Phoradendron flavescens*), although Spanish moss is not extremely abundant except along the outer coastal plain.

Hardwood Cane Bottoms

Alluvial bottomlands that are flooded less frequently than those of mixed hardwoods often develop an association where beech (*Fagus grandifolia*), white oak (*Quercus alba*), and shagbark hickory (*Carya ovata*) become more important as canopy species and may become locally dominant. Under-

story is much the same as in other hardwood bottoms with holly (*Ilex opaca*) possibly becoming more abundant and individual specimens becoming larger. The main herbaceous characteristic is the presence of dense stands of cane (*Arundinaria gigantea*). In some areas cane forms such dense stands as to preclude most other herbaceous growth.

Even though hardwood-cane bottoms seem relatively stable, hardwoods usually succeed cane after disturbances other than fire, in which case cane succeeds hardwoods.

FIGURE 71. Many pines are capable of sprouting new shoots from adventitious buds. This growth was stimulated by a relatively hot fire that burned the shrub layer between the pines. If pinelands are frequently burned, low-intensity fires result, and damage to trees is not as extensive, yet undergrowth is effectively removed.

FIGURE 72. A pond cypress savannah. Very wet conditions often produce picturesque savannahs dominated by flat-topped pond cypress (*Taxodium ascendens*). Shrubby vegetation is usually absent, but herbaceous vegetation is extremely abundant and diversified and is dominated by sedges, rushes, and a multitude of grasses and showy flowering species.

Savannahs

For many years it has been the practice of landowners in the coastal plain to burn their land periodically to prevent the development of thick growths of shrubs and scrub oaks and to promote better bird hunting. This practice, along with frequent naturally occurring fires in the vast pinelands, has created areas of widely spaced trees with little or no understory and an abundance of grasses and herbaceous plants.

The frequent fires seem to maintain most savannahs in a longleaf pine sub-climax, or pyric climax if you wish. These areas may be dominated in the herbaceous layer by toothache grass (*Ctenium aromaticum*), but they also support a great number of other species such as wiregrass (*Aristida stricta*), broomsedge (*Andropogon* spp.), and panic grass (*Panicum* spp.). Most of these grasses are extremely inflammable and contribute significantly to the pyric cycle. Their inflammability aids first of all in development of a mineral seed bed, which in turn aids in the establishment of longleaf pine seedlings. Second, ground fires aid in removing competing vegetation of longleaf pines. This is extremely effective except during the first year and again during the sixth to eighth years of growth when the pine itself is susceptible to fire. Lastly, fires evidently aid in reducing damage from brown spot, a common disease of longleaf pines.

FIGURE 73. If protected from fire during periods of low moisture, the pond cypress savannah may be invaded by various hardwoods such as gum (*Nyssa sylvatica* var. *biflora*), sweetgum (*Liquidambar styraciflua*), and red bay (*Persea borbonia*), as well as a thick pocosin-type shrub layer.

Soils usually contain a dark humic layer with high organic content; in addition, they may often overlie a clay hardpan. Drainage in most cases is good, but the high water table keeps the surface nearly saturated, although not to the extent where peat develops.

Several types of savannahs exist within the State, each resulting from variances in moisture and fire conditions. In many places it is even possible to find a mosaic of communities, with each dominated by characteristic species reflecting local environmental conditions. These local conditions produce several arborescent dominants, such as pond cypress (*Taxodium ascendens*), longleaf pine, pond pine (*P. serotina*), slash pine (*P. elliottii*), and various oaks including willow oak (*Quercus phellos*), laurel oak (*Q. laurifolia*), water oak (*Q. nigra*), southern red oak (*Q. falcata*), and post oak (*Q. stellata*). Other hardwoods may appear, but their occurrence is generally spotty.

Very wet conditions may produce a picturesque floral pattern dominated by flat-topped pond cypress, a landscape somewhat resembling sections of the Everglades. Pond cypress is usually the only tree species present. Shrubby vegetation is normally absent, but herbaceous vegetation is extremely abundant and diversified. Sedges, rushes, and grasses are most common, but it is the showy flowering species that seem to draw the most interest. Among them are pipewort (*Eriocaulon decangulare*), yellow-eyed grass (*Xyris ambigua* and *X. jupicai*), sabatia (*Sabatia difformis*), blue flag (*Iris virginica*), milkweed (*Asclepias lanceolata*), two species of yellow flowering polygala (a short one, *Polygala ramosa,* and one usually over four decimeters tall, *P. cymosa*), and deer's tongue (*Trilisa odoratissima*). Various species of insectivorous plants are present, with the principal ones being pitcher plant or trumpets (*Sarracenia flava*), hooded pitcher plant (*S. minor*), several species of bladderwort (*Utricularia* spp.), a couple of species of sundew (principally *Drosera leucantha* and *D. capillaris*), and butterwort (*Pinguicula lutea* and a blue flowered species *P. caerulea*).

Orchids are relatively common also, with the most abundant ones being yellow fringed orchid (*Habenaria ciliaris*), crested fringed orchid (*H. cristata*), snowy orchid (*H. nivea*), and grass-pink (*Calopogon pulchellus*). Spreading pogonia or rosebud orchid (*Cleistes divaricata*) is often present, but it is not as common.

Where topography and drainage are similar to the conditions of pocosins but the fire frequency is higher, a typical pine-toothache grass savannah develops. In most areas longleaf, slash, and loblolly pines predominate, but scattered pond pine and pond cypress do appear. Toothache grass (*Ctenium aromaticum*) is the characteristic graminoid in the dense ground cover, but other graminoids such as wiregrass, colorful rush (*Dichromena latifolia*), broomsedge (*Andropo-*

FIGURE 74. A pine-toothache grass savannah. In addition to the obvious widely spaced stand of pine, the herbaceous layer is dominated by a mixture of grasses of which toothache grass (*Ctenium aromaticum*) is a major constituent. The shrub layer is usually held in check by periodic fires, and although most of the area is relatively moist, thistles and other drier-site plants are found on anthills and other mounds.

gon spp.), muhly (*Muhlenbergia expansa*), umbrella grass (*Fuirena squarrosa*), and beak rush (*Rhynchospora* spp.) appear very frequently.

Common forbs include colicroot (*Aletris aurea* and *A. farinosa*), several asters (*Aster dumosus, A. paludosus* and *A. squarrosus*), red root (*Lachnanthes caroliniana*), marshallia (*Marshallia graminifolia*), meadow beauty (*Rhexia alifanus*), lobelia (*Lobelia nuttallii*), and rayless goldenrod (*Chondrophora nudata*). The usual assortment of insectivorous plants are present as well as a few members of the Orchidaceae.

Pine-toothache savannahs often intergrade with pine-wiregrass dominated sites. In most cases this continuum is brought about by somewhat better drainage and the possible occurrence of an underlying hardpan.

Oak savannahs have basically the same edaphic conditions as do pine savannahs and probably result from extremely aggressive oak transgression coupled with an erratic fire cycle. Typical quercine species include willow oak, southern red oak, laurel oak, water oak and post oak. Naturally the expected pine species are still present, although in lesser quantities than in pine savannahs. Ground cover also is similar to that of the pine savannah. Shrubs also increase in perspective as a result of the erratic fire cycle. Species vary depending on location and moisture availability, but nevertheless, typical ones include wax myrtle (*Myrica cerifera*), inkberry (*Ilex glabra*), persimmon (*Diospyros virginiana*), and winged sumac (*Rhus copallina*).

As mentioned before, coastal plain savannahs are fire maintained, and if protected from fire, probably would succeed to the vegetation type most characteristic of the local edaphic conditions. In light of this, examination of successional stages shows that pine-toothache savannahs usually would progress to a bay forest or pocosin habitat, depending

FIGURE 75. A pine-wiregrass savannah. These savannahs are somewhat better drained than pine-toothache grass savannahs, and may have several individual herbaceous unions, all of which have wiregrass (*Aristida* spp.) as a dominant constituent. If protected from fire for an extended period, they may succeed to an oak-hickory dominated forest type.

upon the water table and drainage; whereas drier sites of pine-wiregrass and oak savannahs would most probably culminate in an oak-hickory forest, after going through stages resembling mesic hardwood and pine habitats.

Lotic Communities

Major streams and rivers differ from standing water communities in three criteria: current, oxygen content, and amount of land-water interaction. There are also noticeable differences between major streams and large rivers such as the Pee Dee, Santee, and Edisto that carry a very high volume of water.

Botanically speaking, sluggish rivers and streams with their bordering pools and eddies are most productive. In these areas large accumulations of silt and other substrates allow for attachment and rooting of several common genera. Among these are water-weed (*Elodea canadensis*), tapegrass (*Vallisneria americana*), pondweed (*Potamogeton* spp.), and bushy pondweed or water nymph (*Najas guadalupensis*). Emergent species include bur-reed (*Sparganium americanum*), arrowhead or duck potato (*Sagittaria* spp.) and in some cases cattail (*Typha latifolia*). A second species of cattail, *T. glauca*, is known to occur in shallow waters of fresh or slightly brackish lakes, ponds, and rivers. In the latter species the staminate and pistillate flowers are usually separated slightly and pith at the stem base is somewhat off-white to buff colored. Permanent attachment of plants in swift currents and in shoals is difficult for most species. In these areas algae such as *Cladophora* and *Spirogyra* predominate, and along with the algae occur various species of diatoms and aquatic mosses such as *Fontinalis*.

As mentioned before, land-water interaction is extremely important in providing both substrate and nutrition for aquatic species. The greater this interchange, the more luxuriant the growth of primary producers.

FIGURE 76. Aggressive oak transgression or erratic fire maintenance in flat lands of the coastal plain often results in an occasional open stand of dominant oaks known as an oak savannah. The most prevalent trees are willow oak (*Quercus phellos*), laurel oak (*Q. laurifolia*), black oak (*Q. velutina*), post oak (*Q. stellata*), and southern red oak (*Q. falcata*), plus the usual pines such as longleaf (*Pinus palustris*). Ground cover is similar to that of a pine savannah.

Limestone Sinks

The middle and outer coastal plain is underlaid by large deposits of Santee limestone dating all the way to the middle Eocene epoch, some 50 to 60 million years ago. In many areas limestone is close enough to the surface to have produced sinks, jamas, and small caves through a gradual weathering process. Several large concentrations of sinks and subterranean passages may be found in various counties. One such concentration is in an area near Santee State Park. Other large and well-developed sinks may be found in Barnwell, Williamsburg, Horry and Berkeley counties. There are undoubtedly other outcroppings, but their extent and development are not as widely known.

As mentioned before, these limestone dissolution formations may take on several aspects. The simplest (or least eroded) form is the typical lime sink that harbors a variety of vegetation depending upon the amount of water and basic ions present. The vegetational cover may also take on several aspects.

One limestone sink near Williston, Barnwell County, is nearly oval depression filled with water. The majority of this area is devoid of trees, but it does support a tremendous growth of water lilies (*Nymphaea odorata*). The border of the sink is much more interesting from a vegetational standpoint than its body. There is very little emergent vegetation present, and as a result, floating vegetation often reaches the shrub zone. Typical shrubs include ti-ti (*Cyrilla racemiflora*), myrtle (*Myrica cerifera*), swamp honeysuckle or swamp azalea (*Rhododendron viscosum*), deerberry (*Vaccinium stamineum*), and dangleberry (*Gaylussacia frondosa*). There is also an abundance of greenbrier (*Smilax laurifolia*) and wild grape (*Vitis* spp.).

FIGURE 77. Ditch Pond, Barnwell County, is apparently a lime sink formed in the Aiken Plateau and has a conspicuous absence of typical pool zonation (*Radford, 1974*). There is however, a well-developed shrub community in the background. The most striking vegetation is the abundance of water lilies (*Nymphaea odorata*), but there are also several relatively rare herbaceous species found along the margin.

Understory species and small trees farther from the center of the pond include red bay (*Persea borbonia*), red maple (*Acer rubrum*), and blackgum (*Nyssa biflora*). Along the inner edge of the shrubby vegetation may be found several herbaceous species of interest. Two species of bladderwort (*Utricularia olivacea* and *U. biflora*) grow in relative abundance, as well as pitcher plants and sundews where a peat layer has formed.

Other sinks exist near Honey Hill, Berkeley County, and near Murrells Inlet although the sinks are actually in Horry County. These sinks are much smaller and support a thick growth of pond cypress and tupelo (*Nyssa sylvatica* var. *biflora*). Around most of the ponds there is an area where deep shade and fluctuating water levels have kept most vegetation out, except for pond spice (*Litsea aestivalis*) which grows in relative abundance. Pond spice is one of South Carolina's rarest shrubs and generally grows best in and around lime sinks. The remainder of the vegetation, except for a few herbaceous species that occur principally near lime sinks, is characteristic of most pocosins and swamp forests.

A very intensive development of small sinks, jamas, and caves may be seen from Eutawville to the eastern and southwestern boundaries of Orangeburg County. One high concentration of these is found near Santee State Park where underground streams emerge from limestone caves and drain into Lake Marion. According to Siple (1960, 1967, 1975) action of the Coriolis force has accentuated the solution process in the subsurfaces, producing tubular openings and caves through which ground water is discharged into the main stream and its tributaries. As would be expected, there is a very lush herbaceous and shrubby growth at the mouths of these streams. The surrounding area is generally a mesic pine-mixed hardwood community, and as is common in an area influenced by limestone deposits, certain species such as

FIGURE 78. One of South Carolina's rarest shrubs, pond spice (*Litsea aestivalis*), appears in the left foreground behind the cane. It is found in several lime sinks and bays in several counties of the coastal plain. Pond spice is usually associated with red bay (*Persea borbonia*), red maple (*Acer rubrum*), bald cypress (*Taxodium distichum*), and sweet bay (*Magnolia virginiana*), as well as numerous other plants suited to this wet habitat.

redbud or Judas tree (*Cercis canadensis*) appear rather frequently.

Old Growth Pine Communities

Old growth pine communities occur over a wide variety of submesic and subxeric sites of the piedmont and coastal plain. In most cases, these pine communities develop after such disturbances as fire or clear cutting in hardwood or mixed forests; however, old field succession may also lead to similar communities under ideal circumstances. In addition, continued fire protection leads to very thick growths of pine on piedmont sites where conditions are not conducive to hardwood development.

Such old growth pine normally culminates in a closed canopy with very sparsely populated understory, herb, and shrub layers. The majority of understory species are seedlings or saplings of various hardwood species. The floor of the forest is usually covered with copious amounts of needle litter which, because of its relative acidity, somewhat retards hardwood seed germination in some species; moreover, further invasion of hardwood species is often checked by occasional fires. Because of the distribution of these pine communities, there are typically three species that dominate: shortleaf pine (*Pinus echinata*) in the upper piedmont, loblolly pine (*P. taeda*) in the lower piedmont and inner coastal plain, and longleaf pine (*P. palustris*) over the majority of the remainder of the coastal plain.

It is more common for coastal plain pine communities to be maintained in their homogeneous state by periodic fires than piedmont forests, and as a result, coastal plain forests tend to have a more open canopy and a correspondingly lower incidence of seedlings, both of pine and hardwood species.

In coastal plain stands where hardwood introduction has been successful, some of the most common hardwoods are sweetgum (*Liquidambar styraciflua*), persimmon (*Diospyros virginiana*), pignut hickory (*Carya glabra*), and mockernut hickory (*C. tomentosa*). Quercine species are numerous, generally depending upon local moisture supply. Most prevalent in subxeric habitats are bluejack oak (*Q. incana*), blackjack (*Q. marilandica*), turkey oak (*Q. laevis*), and post oak (*Q. stellata*). Submesic species include black oak (*Q. velutina*), southern red oak (*Q. falcata*), laurel oak (*Q. laurifolia*), and live oak (*Q. virginiana*).

As mentioned before, shrubs, herbs, and lianas are sparse. Most commonly encountered shrubs are beauty-bush (*Callicarpa americana*), dwarf huckleberry (*Gaylussacia dumosa*), and inkberry (*Ilex glabra*). Lianas which predominate are trumpet vine (*Campsis radicans*) and yellow jessamine (*Gelsemium sempervirens*).

Herbaceous species are more varied, although by no means do they blanket the forest floor with color. Subxeric predominates include two legumes, lupine (*Lupinus villosus*) and sensitive brier (*Shrankia microphylla*), and composites, elephant's foot (principally *Elephantopus nudatus* and *E. elatus* along with *E. carolinianus* and *E. tomentosus*) and deer's tongue (*Trilisa odoratissima*). Also present are tread-softly (*Cnidoscolus stimulosus*) and sticky foxglove (*Aureolaria pectinata*). Submesic species include two blue flowered violets (*Viola septemloba* and *V. walteri*), black-root (*Pterocaulon pycnostachyum*), spotted wintergreen (*Chimphila maculata*), and blue lobelia (*Lobelia nuttallii*). Ebony spleenwort (*Asplenium platyneuron*) is a frequent pteridophyte, but of course, bracken fern (*Pteridium aquilinum*) is omnipresent.

Coastal Dunes

A cross-section of mature dunes show several well-developed physiographic zones: a fore dune, a depression immediately behind it, and a back dune; another depression is often found behind the back dune. Behind the dunes lie extensive salt marshes having vegetation directly affected by amounts of flooding and relative salt concentration.

Progressing landward, we find a narrow shrub zone, then a true maritime forest. Both zones show direct correlation to salt spray intensity and elevation.

Sea oats (*Uniola paniculata*) are dominant at nearly all dune sites, especially on exposed sites of the seaward side and top of the fore dune, and top of the second dune. Marsh elder (*Iva imbricata*) also favors exposed sites and is most common at sites on the front and top of the first dune. Low cord grass or marshhay cordgrass (*Spartina patens*) favors depressions behind dunes but may also be found on the front and top of the fore dune. It is also a dominant constituent of the marsh border zonation. Sandspurs (*Cenchrus tribuloides*) are usually found in sheltered depressions along with dune spurge (*Euphorbia polygonifolia*), dune pennywort (*Hydrocotyle bonariensis*), poor-Joe (*Diodia teres*), daisy fleabane (*Erigeron canadensis*) and golden aster (*Heterotheca subaxillaris*). It may also be well to note that not all species found in dune communities are restricted to these habitats. Some species such as daisy fleabane, golden aster, and poor-Joe are very widespread in dry habitats throughout the state. Moreover, several individuals of water spider orchid (*Habenaria repens*) have been found in a large depression behind a fore dune in Georgetown County (Barry 1968). These plants are obviously well out of their normal freshwater habitat, further illustrating their extremely broad ecological tolerance.

FIGURE 79. These coastal dunes, as well as many others of the coast, are stabilized by plants such as sea oats (*Uniola paniculata*), sandspurs (*Cenchrus tribuloides*), and other species well adapted to a high salt spray environment. In addition, many species, such as several specimens of water spider orchid (*Habenaria repens*), found in the blowout depression in the foreground, are able to survive if afforded some protection from the salt spray.

In addition to the previously mentioned dominants, many species may be classified as dune codominants. These include grasses such as broomsedge (*Andropogon virginicus*), dune panic grass (*Panicum amarum*), beach grass (*Ammophila breviligulata*) and sand grass (*Triplasis purpurea*), sedges (*Cyperus* spp.), and several other species such as evening primrose (*Oenothera laciniata*), wild bean (*Strophostyles helvola*) and croton (*Croton punctatus*).

Various studies have been made on aspects of salt spray tolerance, but in general, soil salinity and physical conditions affecting water-holding capacity have been considered as two of the most important determinants of distribution in strand species. Precipitation/evaporation influences are usually too constant to be of value; however, temperature and light factors are widely variable. Stalter (1974a) found that root penetration of dune species was shallow for all species; most species tested having roots located at depths of 2.9 inches to 14.9 inches (7.5 to 38 cm). Few roots were found beyond 23.6 inches (60 cm).

Dune-Shrub Communities

At the rear of the dune exists a shrub community composed principally of salt spray resistant species. With the exceptions of Hercules' club (*Zanthoxylum clava-herculis*) and marsh elder (*Iva frutescens*), which are usually restricted to dune communities, most of the other species are also found in other areas of the coastal plain and piedmont. Most predominant in the dune shrub community are groundsel (*Baccharis halimifolia* and *B. angustifolia*), several vines such as wild grape (*Vitis rotundifolia*), Virginia creeper (*Parthenocissus quinquefolia*), and greenbrier (*Smilax* spp.). Other shrubby species present are yaupon holly (*Ilex vomitoria*), wax myrtle (*Myrica cerifera*), winged sumac (*Rhus copallina*), and poison oak (*R. toxicodendron*). In certain instances,

FIGURE 80. A dune community dominated by sea oats (*Uniola paniculata*). Between the dunes and the land mass there is a narrow estuary dominated by cordgrass (*Spartina alterniflora*). The wind direction causes salt spray to be blown inland, causing a sheared effect to the shrubs and trees in the background.

scrubby forms of live oak (*Quercus virginiana*) and southern red cedar (*Juniperus silicicola*) may be found. Southern red cedar differs from eastern red cedar in that the former has more slender twigs and is usually restricted to the outer coastal plain and maritime communities, whereas eastern red cedar grows over the remainder of the state in suitable habitats.

In a study of dune vegetation in South Carolina, Rayner and Batson (1976) also found that there was a distinct shrub-vine zone immediately landward of the dune community and that this shrub-vine zone usually became narrower as more protection is afforded by open dunes. It was also observed that species diversity increases as salt spray intensity decreases.

Maritime Forest

In communities bordering expanses of salt water, salt spray is enough to influence the development of a climax vegetation quite different from areas more removed. This phenomenon has led many researchers to consider these areas to be a specific type of climax community, namely, the salt spray climax (Wells, 1939; Boyce, 1954).

Asymmetrical growth of these coastal shrubs and trees (especially live oak) may be explained in several possible ways. Primarily, the constant wind exposure and an excess transpiration tends to cause dessication of young twigs, an effect also noted at high mountain elevations; moreover, a sandblasting effect by wind carried soil and salt particles also tends to harm tender shoots. This combination of wind and salt spray tends to be the most plausible explanation for the observed pruning effect. Boyce (1954) even suggests correlations between the angle of tree canopy and the degree of exposure to wind velocity. He found that coastal wind velocities tend to range from a low of 3.3 to 9.9 ft/sec (1 to 3 m/sec) to a high of 59.2 to 65.8 ft/sec (18 to 20 m/sec). During non-storm periods velocities follow a diurnal pattern with nights usually being calm and wind rising to a peak between 2:00 and 4:00 P.M.

The genetic tolerance of certain species to this salt dessication is much greater than others. *Quercus virginiana*, with its high tolerance to salt spray, is seldom so completely dominant away from the maritime zone. In addition to *Quercus virginiana*, other trees having a relatively high salt spray tolerance are palmetto (*Sabal palmetto*), dwarf palmetto (*S. minor*), and slash pine (*Pinus elliottii*) (Penfound and O'Neill, 1934; and Penfound and Burleigh, 1941). This variance in tolerance is used to delimit true maritime forests, and also to explain the zonation that occurs.

Inward from the shrub zone occupied by groundsel (*Baccharis halimifolia*) there is a steady increase of new taxa corresponding to a decreasing order of resistance to salt spray. Loblolly pine (*P. taeda*) and turkey oak (*Q. laevis*), which are common on dry sandy sites inland, are strongly intolerant; therefore, their appearance would tend to delimit the inner boundary of the true maritime forest (Wells and Shunk, 1937, 1938), nevertheless, the wide genetic tolerances of loblolly and slash pines (*P. elliottii*) allow species to appear in sheltered areas of the maritime forest.

Other arborescent species found in association with the maritime species are red bay (*Persea borbonia*), wild olive (*Osmanthus americana*), and to some extent laurel oak (*Q. laurifolia*). Associated shrubs include beauty-bush (*Callicarpa americana*), wax myrtle (*Myrica cerifera*), yaupon holly (*Ilex vomitoria*), Hercules' club (*Zanthoxylum clava-herculis*), and winged sumac (*Rhus copallina*). Principal lianas include yellow jessamine (*Gelsemium sempervirens*), greenbrier (*Smilax* spp.), Virginia creeper (*Parthenocissus quinquefolia*), poison ivy (*Rhus radicans*) and pepper-vine (*Ampelopsis arborea*).

FIGURE 81. Hammock vegetation on Turtle Island, Jasper County. In looking from the low marsh dominated by cordgrass (*Spartina alterniflora*) toward a border of black rush (*Juncus roemerianus*) and various brackish marsh grasses, it is evident that a small increase in elevation has caused a decrease in the amount of salt present. Beyond this is a narrow shrub zone followed by hammock vegetation dominated by palmetto (*Sabal palmetto*), live oak (*Quercus virginiana*), pines (*Pinus taeda*), and wild olive (*Osmanthus americana*). The dense interior of the hammock has areas of vines and shrubs that make it almost impenetrable.

FIGURE 82. Late evening along a salt marsh with a dominance of cordgrass (*Spartina alterniflora*) all the way to the maritime forest. A large live oak is in the foreground (*courtesy S.C.W.M.R. Department*).

Characteristic grasses are dallis grass, (*Paspalum* spp.), panic grass (*Panicum* spp.), bermuda grass (*Cyndon dactylon*), finger grass (*Chloris petraea*), St. Augustine grass (*Stenotaphrum secundatum*), and sandspurs (*Cenchrus tribuloides*). Dominant forbs include poor-Joe (*Diodia teres*) and Florida clover (*Richardia scabra*) with seashore mallow (*Kosteletskya virginica*), knotweed (*Polygonum* spp.), sedges (*Cyperus* spp.), and rushes (*Juncus* spp.) being locally abundant where conditions permit.

One of our rarest shrubs, a plant called shell mound shrub (*Sageretia minutiflora*), is found nearly exclusively on old shell mounds. In these areas calcareous shells promote a luxuriant growth of southern red cedar (*Juniperus silicicola*) and Spanish bayonet (*Yucca aloifolia*). Some areas are rendered almost impenetrable by these plants after they have attained considerable age. Associated shell mound species are very much the same as in maritime zones.

Salt Marsh and Border Zonation

Some of the least frequently visited areas along the coast of South Carolina are the broad expanses of salt marshes. These often extend continuously for miles and are often more than a mile wide. Viewed from the air, drainage patterns of marshes give strikingly different appearances. A dendritic pattern often characterizes the softer areas of the marsh; whereas deep channels predominate in firmer areas.

Upon closer inspection, most marshes include four vegetation zones. Each zone is classified by its relative elevation above datum (mean low tide level), and each is found to support distinctive species.

The upper high marsh is dominated by marsh elder (*Iva frutescens*), sea myrtle or groundsel (*Baccharis halimifolia* or *B. angustifolia*), and marshhay cordgrass (*Spartina patens*). It is flooded on the average less than an hour a day and has low

salt concentration of from 0.1 to 0.5 percent. The low high marsh is dominated by glasswort (*Salicornia virginica*), smooth cordgrass (*Spartina alterniflora*), and sea oxeye (*Borrichia frutescens*), with sea lavender (*Limonium carolinianum*) and saltgrass (*Distichlis spicata*) as associates. This level is usually flooded daily and has a higher salt concentration (0.3 to 3.0 percent).

Upper low marshes are dominated by a pure stands of dwarf cordgrass and are usually flooded twice daily; whereas the lower low marsh is dominated by tall cord grass and is flooded on the average of 14 hours per day at its lowest portion. Salt concentrations for the upper and lower levels of the low salt marsh are 0.5 to 3.2 percent (and up to 5.0 percent in salt pans) and 0.5 to 3.0 percent respectively (Stalter, 1973b). Most authors agree that the depth and duration of flooding of the marsh controls the characteristic zonation of marsh vegetation.

In examining the tolerance of various marsh species, Stalter (1973 a, b) found that marshhay cordgrass was least tolerant to long periods of flooding. He also found that sea oxeye was found to be more tolerant than marshhay cordgrass but less tolerant than glasswort, sea lavender, and cordgrass; however, none except cordgrass was able to endure long periods of flooding.

It also might be noted that the genetic relationship between dwarf and tall cordgrass is still not settled. Stalter and Batson (1969) postulated that two forms of cordgrass may exist, but equally important are the results of Mooring et al. (1971) that suggest that height differences found in North Carolina marshes might best be defined as ecophenes. In South Carolina marshes it is possible to find cordgrass growing immediately around small potholes somewhat taller than that growing only a foot or so away from the small body of water. This suggests that the amount of available water and

FIGURE 83. Zonation along the salt marsh border is extremely noticeable in this setting. There is a definite change from a marsh dominated by cordgrass (*Spartina alterniflora*) to elevations successively dominated by black rush (*Juncus roemerianus*) various grass (*Andropogon virginicus*, *Panicum* spp., *Setaria geniculata*, and *Cenchrus tribuloides*), marsh elder (*Iva frutescens*), and other shrubs such as wax myrtle (*Myrica cerifera*) and yaupon holly (*Ilex vomitoria*).

the immediate salt concentration might be a limiting factor in the height of the plant.

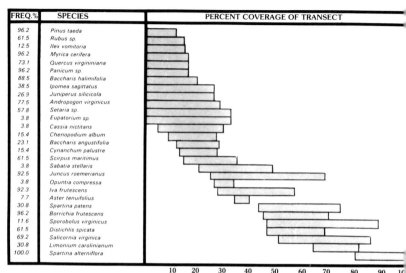

FREQ.%	SPECIES	PERCENT COVERAGE OF TRANSECT
96.2	Pinus taeda	
61.5	Rubus sp.	
12.5	Ilex vomitoria	
96.2	Myrica cerifera	
73.1	Quercus virgininiana	
96.2	Panicum sp.	
88.5	Baccharis halimifolia	
38.5	Ipomea sagittatus	
26.9	Juniperus silicicola	
77.5	Andropogon virginicus	
57.8	Setaria sp.	
3.8	Eupatorium sp.	
3.8	Cassia nictitans	
15.4	Chenopodium album	
23.1	Baccharis angustifolia	
15.4	Cynanchum palustre	
61.5	Scirpus maritimus	
3.8	Sabatia stellaris	
92.5	Juncus roemerianus	
3.8	Opuntia compressa	
92.3	Iva frutescens	
7.7	Aster tenuifolius	
30.8	Spartina patens	
96.2	Borrichia frutescens	
11.6	Sporobolus virginicus	
61.5	Distichlis spicata	
69.2	Salicornia virginica	
30.8	Limonium carolinianum	
100.0	Spartina alterniflora	

10 20 30 40 50 60 70 80 90 10

FIGURE 84. Zonation of species on salt marsh borders as compiled from 26 transects in Georgetown County (*Barry, 1968*).

FIGURE 85. A typical brackish marsh dominated by black rush (*Juncus roemerianus*).

Brackish Marshes

Marshes dominated by cordgrass are usually considered to have a salinity above 15,000 ppm (Soils Memorandum SC-4, 1962). If the salinity content drops below 15,000 ppm, as in areas fed partially by freshwater or only periodically flooded by saltwater, dominant vegetation is black rush (*Juncus roemerianus*).

On higher areas flooded only by full moon or storm tides, most common species are glasswort, seashore saltgrass (*Distichlis spicata*), and sea oxeye. Often there is freshwater drainage from the mainland that supports an even more marked effect on the zonation along the marsh border.

Major river deltas such as the Cooper River in Charleston County also have a zonational change from purely salt marshes to freshwater marshes with an intermediate brackish zone. Stalter (1973 b) characterized the Cooper River estuary as having the above three zones, each with its unique floristic composition. The most important parameters in this case seem to be salinity and duration and depth of flooding, although a combination of these factors and relative acidity give widely varing ecological conditions.

Freshwater Marshes

Freshwater marshes are wetlands where the dominant vegetation consists of rushes, sedges, grasses, cattails, and small hydrophytic angiosperms. Biologically, marshes are among the most interesting and diverse communities in the State, because of their high diversity of species composition and ecological conditions.

Freshwater marshes are most extensively developed along rivers subject to tidal influence, such as the upper Cooper River estuary. In addition, marshes may also be created both naturally and artificially along level pond and lake margins and in managed impoundments.

In classifying coastal plain marshes it is useful to place them into two categories: inland fresh marshes and coastal fresh marshes. Coastal fresh marshes such as the upper reaches of estuaries have a flora essentially the same as marshes not influenced by tidal action, yet fresh marshes in coastal areas to have some species that do not occur farther inland. Inland marshes have a water-logged substrate during the growing season which is often covered by six inches (15.2 cm) or more of water. Predominant plants are naturally hydrophytes, and most are herbaceous or small shrubs. Since most of the aquatics encountered are emergents, they are quite flexible and nearly all develop tough, fibrous rhizomes

SELECTED SPECIES AND THEIR DISTRIBUTION ALONG THE COOPER RIVER ESTUARY																						
Taxa	**Stations**																					
	1	2	3	4	5	6	7	8	9	10	11	12	13	14	15	16	17	18	19	20	21	22
Saururus cernuus	•	•	•	•	•																	
Pontederia cordata	•	•	•	•	•	•	•			•	•											
Zizaniopsis miliacea	•	•	•	•	•	•	•	•	•	•	•	•		•								
Cicuta maculata		•	•			•	•	•	•	•	•	•	•	•								
Spartina cynosuroides			•	•			•	•	•		•	•	•	•	•	•						
Cladium jamaicense				•	•	•	•	•	•	•	•	•	•	•		•						
Juncus roemerianus						•	•	•	•	•	•	•	•	•	•	•	•	•				
Liliaeopsis chinensis						•		•		•	•	•			•	•			•			
Solidago sempervirens									•					•	•	•			•			
Eleocharis albida										•						•						
Spartina alterniflora															•	•	•	•	•			
Borrichia frutescens																•	•	•	•			
Iva frutescens															•	•	•	•	•			•
Spartina patens																•	•	•				•
Scirpus robustus																•		•				
Atriplex patula																•	•					•
Salicornia virginica																•			•	•		•
Baccharis halimifolia																•			•	•		•

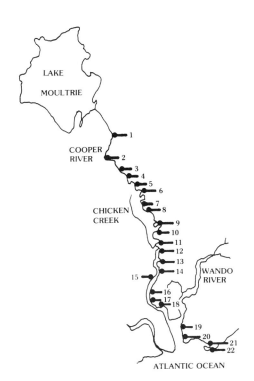

FIGURE 86. Estuarine sampling stations along the Cooper River in Charleston and Berkeley counties and significant species associated with each. Stations 1-5 are freshwater, 6-14 are brackish, and 15-22 are saline (*modified from Stalter, 1974b*).

and roots. These often develop into a firm mat, resisting both wind and water dislocation. Such a development is similar in many aspects to floating vegetation mats in inland lakes, swamps, and old millponds.

Dominants include various species of spike-rush (*Eleocharis* spp.), bulrush (*Scirpus validus*), bur-reed (*Sparganium americanum*), cattails (*Typha latifolia* and *T. angustifolia*), panicum (especially *Panicum hemitomon*), and redroot (*Lachnanthes tinctoria*). Giant foxtail (*Setaria magna*) and pickerelweed (*Pontederia cordata*) are locally abundant, and at times can offer a spectacular display when in bloom. Two species of barnyard grass (*Echinochloa* spp.) are present. *E. crusgalli* is dispersed throughout the state, whereas *E. walteri* is found principally in outer coastal plain marshes.

As mentioned before, most coastal fresh marsh vegetation is very similar to that of inland fresh areas with a few exceptions. Major exceptions are the presence of giant reed (*Phragmites communis*), southern wild rice or water millet (*Zizaniopsis miliacea*), and wild rice (*Zizania aquatica*) in the coastal freshwater and slightly brackish marshes. Two species of cattail (*T. angustifolia* and *T. domingensis*) are found in slightly brackish alkaline marshes and pools.

Glossary

achlorophyllous not having chlorophyll

acidophyllous pertaining to a plant that lives in acid soil (below pH 6.5)

alluvial moved by or formed by water movement

angiosperm a flowering plant; a member of the group of plants characterized by having the ovules enclosed in an ovary

annual a plant in which the entire life cycle is completed in a single growing season

arborescent of treelike habit

association one of at least two units of a climax status, floristically and usually geographically distinct, yet ecologically related

bald a high ridge or steep slope vegetated by densely compact and almost impenetrable shrub communities usually composed of ericaceous species

base level the level below which land surface cannot be reduced by running water

biotic pertaining to living matter and living organisms

buttressed flaring out at the base

calcareous consisting of or containing calcium carbonate

calciphyte plants living on basic soil abundantly supplied with calcium ions

Cambrian a period of the earth's history dating approximately from 500 million to 620 million years ago

canopy within most woody communities the top layer of leafy growth

cellulose a complex carbohydrate formed from glucose; it is the chief component of the cell wall in most plants

centrifugal growing or proceeding in a direction away from the center

chlorophylls green pigments involved in photosynthesis

climax stage the terminal community of a succession, which is maintained with little change provided the environment does not change significantly

community an assemblage of organisms living together

composite any member of the daisy family (*Asteraceae*)

conglomerate rock composed of rounded fragments varying from small pebbles to large boulders in a cement (as of hardened clay)

conifer a member of group of mostly evergreen trees and shrubs (including pines) whose seeds are produced in a cone and not in a true fruit

continuum a series of ecological communities whose vegetation gradually changes along an environmental gradient

Coriolis force an apparent force formed as a result of the earth's rotation which deflects moving objects to the right in the northern hemisphere

cove a deep recess or sheltered area in the side of a mountain

Cretaceous a period in the earth's history approximately 65 million to 135 million years ago when flowering plants were becoming the dominant form of vegetation

crustose crustlike

cutin a fatty substance, rather impermeable to water

cylonic storms a storm center which rotates around a center of low pressure

datum mean low tide level

deciduous referring to plants whose leaves or branchlets fall off in certain seasons

delta an alluvial deposit at the mouth of a river

dendritic a tree-like system of branching

dominance the influence or control over an ecological community exerted by a particular species or group of species

disjunct in the sense of plants, a population of plants found growing very far from its principal population or range

disseminate to spread about those parts of a plant capable of reproducing, e.g., seeds, fruits

ecophene growth forms which vary with ecological factors

ecotone the place where two major communities meet and blend together

edaphic resulting from or pertaining to soil factors

embayment a conformation resembling a bay

emergent vegetation aquatic vegetation that is rooted, yet the top extends above the water surface

endemic restricted to a particular locality or region

ephemeral a plant that grows, flowers, and dies within a very short period of time

epidermis the surface layer of cells on a plant before cork is formed

epiphyte a plant that grows attached to some larger plant, usually on tree trunks or branches, without deriving nourishment from its host

ericaceous having characteristics similar to those of the heath family

ericad a member of the *Ericaceae*, or heath family

escarpment a long cliff or steep slope separating two comparatively level or more gently sloping surfaces and resulting from erosion or faulting

estuaries bodies of water bordered by and partially cut off from the ocean by a land mass that did not originate from the sea; the place where freshwater joins saltwater along the coast

fibrous finely divided; such as the roots of most grasses

fluvial fan a fan shaped delta of a river or river system

foliose leaflike in form

forbs herbaceous plants other than grasses

fruticose shrublike in form

genetic tolerance the amplitude of ecological conditions which a plant can endure as a result of its genetic makeup

geomorphology the science that deals with the land and features of the earth's surface

gneiss a metamorphic rock corresponding in composition to granite

gorge a narrow steep-walled passage between mountains

gradient a stepwise change in characteristics or concentration

graminoid grass-like

gymnosperm any member of the group of plants that includes the pines, cedars and various other needle-leaved plants which produce their seeds in a cone or sometimes surrounded by fleshy tissue, but not an ovary or fruit

hardpan a cemented or compacted and often clayey layer in soil that is impermeable by roots

hardwood the wood of an angiospermous (flowering) tree as opposed to a conifer

headwaters the beginning of a stream or river

heath an area dominated by plants of the heath family (*Ericaceae*)

herbaceous having little or no woody tissue; fleshy

heterophyte a plant that cannot manufacture its own food but must obtain it from some external source

humus a complex mixture of partially decomposed organic matter in the soil

hydric having copious amounts of moisture

igneous formed from volcanic activity

insectivorous referring to a plant which captures and digests insects

insolation solar radiation that has been received

jamas a large steep-sided solution sink with a chimney-like shaft up to 300 feet in depth

leach to dissolve out by percolating water

legume any member of the legume family (*Fabaceae*); the fruit of a legume

lentic referring to an aquatic habitat in which there is still water, e.g., a pond or lake

liana a climbing herbaceous or woody vine

lichen a complex (usually grayish colored) plant made up an association of fungus and algae, e.g., reindeer moss, British soldiers

lignin a complex organic compound frequently associated with cellulose in plant cell walls

littoral zone a shallow water zone around a body of water where light penetrates to the bottom

loam a type of soil whose excellent garden characteristics are the result of a mixture of sand, silt, clay, and organic matter

lotic referring to an aquatic habitat where there is flowing water, e.g., a river or stream

maritime bordering on the ocean

mesophytic having conditions of moderate amounts of moisture

metamorphic rock produced or changed by heat and pressure which usually results in a more compact and more highly crystalline condition

mineralization the process of converting organic material into inorganic minerals

Miocene an epoch in the earth's history approximately 13 million to 25 million years ago when mammals were at the height of their evolution

monadnock a hill or mountain of resistant rock surmounting a peneplain

morphologic pertaining to the study of the structure and form of a plant or animal

nitrification the conversion of ammonia or ammonium compounds to nitrites and then to nitrates by the activity of certain soil bacteria

Oligocene an epoch in the earth's history approximately 25 million to 36 million years ago when there was a maximum spread of forests

oxbow lake a lake formed by a change in the course of a river

peduncle the main stalk of a flower cluster or the stalk of an individual flower

peneplain a land surface of considerable area and slight relief formed by erosion

perennial a plant that lives for more than two years; it usually flowers annually after a period of vegetative growth

petiole the stalk that attaches a leaf to the stem

phenology the study of periodicity in plants as related to climatic events, e.g. time of flowering

photosynthesis the manufacture of food, mainly sugar, from carbon dioxide and water in the presence of chloroplasts utilizing light energy and releasing oxygen

physiography the description of nature's natural phenomena in general

physiologic pertaining to functions and chemical activities of living material

plicate folded, as in a fan

pocosin an area of dense shrub and vine growth usually on a wet, peaty, poorly drained soil

post-climax a climax which is produced by a more favorable climax, usually cooler and/or moister

prairie a large, poorly drained relatively treeless area dominated by grasses

Precambrian relating to the earliest era of geologic history

primary producer a plant, since it is the organism that captures sunlight and uses it to produce organic material

primary succession plant growth beginning on bare rock or in a pond or like habitat where no vegetation has grown before

pteridophyte ferns and their allied plants

pulvinus the swollen base of the petiole, as in the leaves of many legumes

pyric climax a climax maintained by recurrent fires

quercine attributing to the genus *Quercus*, i.e., the oaks

relic botanically, something that has remained from a past time

rhizome a horizontal underground stem

saprophyte a heterophytic organism that obtains its food from nonliving organic matter

savannah a flat area with wide spaced trees, usually dominated by grasses; a savannah is usually better drained than a prairie

scarp a line of cliffs produced by erosion

sere a group of plant communities that successively occupy the same area from the pioneer to a mesic stage

shrub a woody plant that remains low and that produces several shoots or trunks from the base

stratification being arranged at different levels

subcanopy the layer immediately below the canopy in a forest

subclimax a long persisting vegetational stage immediately preceding the climax

submerged anchored aquatics plants growing in water and attached to the bottom, as opposed to those free floating

substrate the substance acted upon; the soil upon which a plant lives

succulent a plant having fleshy tissue designed to conserve moisture

suffrutescent diminutively shrubby, woody at the base with annual shoots produced

taproot a root which is larger and more conspicuous than the others of the plant, such as a carrot

talus a slope formed by an accumulation of rock debris

terrestrial living on dry land, as opposed to aquatic

topography the configuration of a surface including its relief and the position of its natural features

tortifoliate a plant that is capable of reorienting its leaves

transpiration the loss of water in vapor form from a plant

trifoliolate having three leaflets on a single petiole, e.g., the leaf of poison ivy

unconsolidated not stratified or arranged in layers

understory the plants of a forest undergrowth

xeric having a scant water supply

zonation the arrangement or distribution of kinds of organisms into biogeographic zones

References and Bibliography

Anon., 1975. *South Carolina Heritage Trust Program Final Report Document IV, Elements of Diversity.* Nature Conservancy, Arlington, Va.

Anon., 1975. *South Carolina Historic Preservation Plan for Fiscal Year 1975.* Vol. III, S.C. Dept. of Archives and History.

Au, Shu-fun, 1969. *Vegetation and Ecological Processes on Shackleford Bank, N.C.* Ph.D. dissertation, Duke University.

Barry, John M., 1968. *A Survey of the native Vascular Plants of the Baruch Plantation.* M.S. thesis, Univ. of South Carolina.

——— and W. T. Batson, 1969. The Vegetation of the Baruch Plantation, Georgetown, South Carolina, in Relation to Soil Types. *Castanea* 34:71–77.

Batson, W. T., 1972. *Genera of the Southeastern Plants.* Published by the author, Columbia, S.C.

———, *Outline for Plant Distribution from the Carolina Coast to the Peaks of the Southern Appalachians,* Unpublished.

Beck, William W., 1973. Correlation of Pleistocene barrier islands in the lower coastal plain of South Carolina as inferred by heavy minerals. *Geologic Notes.* Vol. 17(3):68–82. S.C. State Development Board, Division of Geology.

Berry, E. W., 1914. The Upper Cretaceous and Eocene Floras of South Carolina and Georgia. *U.S. Geol. Surv. Prof. Paper* 84:1–200.

———, 1916. The flora of the Citronelle Formation. *U.S. Geol. Surv. Prof. Paper* 98.

Billings, W. D., 1938. The structure and development of old field short-leaf pine stands and certain associated physical properties of the soil. *Ecol. Monog.* 8:437–99.

Birkhead, Paul K., 1973. Some flinty crush rock exposures in northwest South Carolina and adjoining areas of North Carolina. *Geol. Notes* 17(1):19–25. S.C. State Development Board, Division of Geology.

Bordeau, P. F., and H. J. Oosting, 1959. The Maritime Live Oak Forests of North Carolina. *Ecology* 40:148–52.

Bormann, F. H., 1953. Factors Determining the Role of Loblolly Pine and Sweetgum in Early Old-field succession in the Piedmont of North Carolina. *Ecol. Monogr.* 23(4).

Boyce, S. G., 1954. The salt spray community. *Ecol. Monogr.* 24:26–67.

Bozeman, J. R., 1971. *A sociologic and geographic study of the sand ridge vegetation in the Coastal Plain of Georgia.* Ph.d. dissertation. Univ. of North Carolina, Chapel Hill.

Braun, E. Lucy, 1935. The undifferentiated deciduous forest climax and the association segregate. *Ecology* 16:514–19.

———, 1941. The differentiation of deciduous forests of the eastern United States. *Ohio Jour. Sci.,* 41:235–41.

———, 1950. *Deciduous Forests of eastern North America.* Philadelphia: The Blakiston Co., 596 pp.

———, 1955. The phytogeography of un-glaciated eastern United States and its interpretation. *Bot. Rev.* 21:297–375.

Buell, M. F., 1939. Peat formation in Carolina Bays. *Bull. Torr. Bot. Club,* 66:483–87.

——— and R. L. Cain, 1943. The successional role of Southern White Cedar, *Chamaecyparis thyoides* in southeastern North Carolina. *Ecol.* 21(1):85–93.

Butler, Robert J., 1971. Structure of the Charlotte belt and adjacent belts in York County, South Carolina. *Geologic Notes* 15(3):49–62. S.C. State Development Board, Division of Geology.

Burbanck, M. P., and R. B. Platt, 1964. Granite outcrop communities in the Piedmont plateau in Georgia. *Ecology* 45:292–306.

198
REFERENCES

Cain, S. A., 1943. The tertiary character of the cove hardwood forests of the Great Smokey Mountains National Park. *Bull. Torr. Bot. Club* 70:213–35.

Cazeau, C. J., 1961. Notes on the geology and structure of Oconee County, South Carolina. S.C. State Development Board, Division of Geology, *Geology Notes,* v. 5, p. 89–91.

———, 1967. Geology and mineral resources of Oconee County, South Carolina. S.C. State Development Board, Division of Geology. *Bull. No.* 34, p. 1–38.

———, and C. Q. Brown, 1963. Guide to the geology of Pickens and Oconee County, South Carolina. S. C. State Development Board, Division of Geology, *Geology Notes,* v. 7(5), p. 31–40.

Chapman, H. H., 1932. Is the longleaf type a climax? *Ecology,* 13:328–34.

Coile, T. S., 1942. Some physical properties of the B horizons of Piedmont soils. *Soil Sci.,* 54:101–3.

———, 1948. Relation of soil characteristics to site index of loblolly and shortleaf pine in the lower Piedmont region of North Carolina. *Duke Univ. School of Forestry Bull.* 13, 78 pp.

Colquhoun, D. J., 1969. *Geomorphology of the lower coastal plain of South Carolina.* S.C. State Development Board, Division of Geology, 36 pp.

Cooke, C. Wythe, 1936. *Geology of the coastal plain of South Carolina.* U.S. geological Survey Bulletin 867, 194 pp.

Cooper, A. W., 1963. A survey of the vegetation of the Toxaway River gorge with some remarks about early botanical explorations and annotated list of the vascular plants of the gorge area. *Jour. Elisha Mitchell Sci. Soc.* 79:1–22.

Cost, Noel D., 1968. *Forest statistics for the southern coastal plain of South Carolina.* Southeastern Forest Exp. Station, U.S. Forest Service Resource Bull. SE-12, 35 pp.

Craddock, G. R., and C. M. Ellerbe, 1966. *Land resource map of South Carolina.* S.C. Agriculture Exp. Station, Clemson, S.C.

Dansereau, P., and F. Segadas-Vianna, 1952. Ecological study of the peat bogs of eastern North America. I. Structure and evolution of vegetation. *Canada Journal Bot.,* 30:490–520.

Dayton, Bruce R., 1966. The relationship of vegetation to Iredell and other piedmont soils in Granville County, N.C. *Jour. Elish. Mitchell Soc.* 82(2):108–18.

Dennis, John V., 1967. *Woody plants of the Congaree forest swamp, South Carolina.* Ecological Studies Leaflet No. 12, Nature Conservancy, Arlington, Va.

Dennis, W. M. and W. T. Batson, 1974. *The floating log and stump communities in the Santee swamp of South Carolina.* Castanea 39:166–70.

Diller, O. D., 1935. The relation of temperature and precipitation to the growth of beech in northern Indiana. *Ecology,* 16:72–81.

Doering, J. A., 1958. Citronelle age problem. *Bull. Amer. Assoc. Petr. Geol.* 42:764–86.

Domin, K., 1923. Is the evolution of the earth's vegetation tending toward a small number of climatic formations? *Acta Bot. Bohemica,* 2:54–60.

Duke, James A., 1961. The psammophytes of the Carolina Fall-line sandhills. *Jour. Elish. Mitchell Sci. Soc.* 77:3–24.

DuMond, David M., 1970. Floristics and vegetational survey of the Chattooga River gorge. *Castanea* 35(4):201–43.

Eleuterius, L. N., 1975. The life history of the salt marsh rush, *Juncus roemerianus. Bull. Torr. Bot. Club* 102(3):135–40.

Fenneman, N. W., 1938. *Physiography of the Eastern United States.* McGraw-Hill, New York.

Flint, R. F., 1947. *Glacial and Pleistocene geology.* Wiley and Sons, N.Y.

Fowells, H. A., 1965. *Silvics of forest trees of the United States.,* Agriculture Handbook No. 271, Forest Serv., U.S. Dept. of Agriculture, Washington, D.C.

Garren, K. H., 1943. Effects of fire on vegetation of the southeastern United States. *Bot. Review* 9:617–54.

Gettmen, R. W., 1974. *A floristic survey of Sumter National Forest—The Andrew Pickens Division.* M.S. thesis, Clemson Univ., Clemson, S.C.

Gleason, H. A., 1917. The structure and development of the plant association. *Bull. Torr. Bot. Club* 43:463–8l.

————, 1926. The individualistic concept of the plant association. *Bull. Torrey Bot. Club* 53:7–26, *American Midl. Nat.*,
21:92–110, 1939.

Glover, Lynn III, and Akhaury Sinha, 1973. The Virginilina
Deformation, a late Precambrian to early Cambrian (?)
orogenic event in the central piedmont of Virginia and
North Carolina. *Am. Jour. Sci.* 273-A:234–51.

Griffin, Villard S., 1967. Folding styles and migmatization
within the Inner Piedmont belt in portions of Anderson,
Oconee and Pickens Counties, South Carolina. *Geol. Notes*
11(3):37–52. S.C. State Development Board, Division of
Geology.

Griffin, V. S., 1969a. Structure of the Inner Piedmont belt
along the Blue Ridge physiographic front of South
Carolina. *Geol. Soc. of America* 1969, Part 4, Southeastern
Section.

————, 1969b. Migmatitic Inner Piedmont belt of northwestern South Carolina. S.C. State Development Board,
Division of Geology, *Geology Notes* 13: 87–103.

————, 1974. *Geology of the Walhalla Quandrangle, Oconee
County, South Carolina.* S.C. State Development Board,
Division of Geology.

Griffitts, W. R., and W. C. Overstreet, 1962. Granitic rocks of
western Carolina Piedmont. *American Journal Sci.,* 250:
777–89.

Haines, W. H. B., 1967. *Forest statistics for the Piedmont of
South Carolina. Southeast.* Forest Exp. Station, U.S. Forest
Service, Resource Bulletin, SE-9, 35 pp.

Hall, T. F., and W. T. Penfound, 1943. Cypress-gum communities in the Blue Girth Swamp near Selma, Alabama.
Ecology, 24:208–17.

Harlow, W. M., and E. S. Harrar, 1968. *Textbook of Denrology.*
McGraw-Hill, New York.

Harper, F., 1958. *The travels of William Bartram.* New Haven:
Yale Univ. Press, 727 pp.

Hartshorn, Gary S., 1972. Vegetation and soil relationships
in southern Beaufort County, N.C. *Jour. Elisha Mitchell Sci.
Soc.* 88:226–41.

Haselton, George M., 1974. Some reconnaissance geomor-
phological observations in northwestern South Carolina
and adjacent North Carolina. *Geologic Notes* 18(4):60–67.
S.C. State Development Board, Division of Geology.

Hatcher, R. D., Jr., 1969a. Structure of the low rank belt of
northwest South Carolina. *Geol. Soc. Amer. Abs.* for 1969,
Part 4, Southeastern Sec. p. 32.

————, 1969b. Stratigraphy of the Brevard-Poor Mountain
Henderson belt of northwest South Carolina. *Geol. Soc.
Amer. Abs.* for 1969, Part 4, Southeastern Section, p. 31.

————, 1969c. Stratigraphy, petrology, and structure of the
low rank belt and part of the Blue Ridge of northwestern-
most South Carolina. S.C. State Development Board,
Division of Geology. *Geology Notes,* 13: 195–41.

————, 1970. Stratigraphy of the Brevard Zone and Poor
Mountain Area, Northwestern South Carolina. *Geol. Soc.
Amer. Bull.,* 81:933–40.

————, 1972. Developmental model for the Southern Appalachians. *Geol. Soc. of Am. Bull.,* 83:2735–60.

————, 1975. Second Penrose Field Conference: The Brevard Zone. *Geology* 3:149–52.

Hendricks, B. A., 1941. Effect of forest litter on soil temperature. *Chronica Botanica,* 6:440–41.

Heron, S. D., 1958. *The stratigraphy of the outcropping basal
Cretaceous Formations between the Neuse River, North
Carolina, and Lynches River, South Carolina.* Ph.D. disserta-
tion, Univ. of North Carolina.

Hocker, H. W., Jr., 1955. *Climatological summaries for selected
stations in and near the southern pine region, 1921–1950.*
Bulletin 56, Southeastern Forest Experiment Station,
Asheville, N.C.

Hodgkins, E. J., 1958. Effects of fire on undergrowth vegetation in upland southern pine forests. *Ecol.* 39:36–46.

Holt, Perry C., ed., 1970. *The distributional history of the biota
of the Southern Appalachians, Part II, Flora.* Virginia
Polytechnic Institute.

Hulse, Robert C., and William H. Kanes, 1972. Modern
shoreline changes, Edisto Island, South Carolina. *Geol.*

Notes 16(4):88–94. S.C. State Development Board, Division of Geology.

Hunt, Charles B., 1967. *Physiography of the United States.* W. H. Freeman and Company.

―――, 1974. *Natural Regions of the United States and Canada.* W. H. Freeman and Company, 725 pp.

Hunt, K. W., 1943. Floating mats on a southeastern coastal plain reservoir. *Torrey Botanical Club Bulletin* 70:481–88.

Hursh, C. R., and F. W. Haasis, 1931. Effects of 1925 summer drought on Southern Appalachian hardwoods. *Ecol.* 12:380–86.

Johnson, Douglas, 1942. *The origin of the Carolina Bays.* Columbia Univ. Press.

Johnson, Henry S., 1964. *Geology in South Carolina.* S.C. State Development Board. Division of Geology.

Johnson, Thomas F. 1970. *Paleoenvironmental analysis and structural petrogenesis of the Carolina Slate Belt near Columbia, South Carolina.* M.S. thesis, Univ. of South Carolina.

Just, T., 1947. Geology and plant distribution. *Ecol. Monog.,* 17:127–37.

Kearney, T. H., 1901. Report on a botanical survey of the Dismal Swamp region. *U.S. Nat. Herb., Contr.* 5:321–585.

Keever, Catharine, 1950. Causes of succession on old fields of the piedmont, North Carolina. *Ecol. Monog.,* 20:229–50.

―――, 1953. Present composition of some stands of the former oak-chestnut forest in the southern Blue Ridge Mountains. *Ecology,* 34:44–54.

Kellogg, C. E., 1936. *Development and significance of the Great Soil Groups of the United States.* U.S. Dept. Agr. Misc. Pub. 229.

Kelley, A. P., 1922. Plant indicators of soil types. *Soil Sci.,* 13:411–23.

Knox, John N., 1974. *A floristic study of Boggs' Rock, a granite gneiss outcrop in Pickens County, South Carolina.* Thesis, Clemson Univ.

Kramer, P. J., 1944. Soil moisture in relation to plant growth. *Bot. Rev.,* 10:525–59.

―――, W. S. Rilery, and T. T. Bannister, 1952. Gas exchange of cypress knees. *Ecology,* 33:117–21.

Kurz, H., and D. Demaree, 1934. Cypress buttresses and knees in relation to water and air. *Ecology,* 15:36–41.

Laessle, A. M., 1958. The origin and successional relationship of sandhill vegetation and sand-pine scrub. *Ecol. Monogr.* 28:361–58.

Landers, H., 1970. *Climate of South Carolina.* Superintendent of Documents, U.S. Government Printing Office, Washington, D.C.

Larson, S. and Wade Batson, 1978. The Vegetation of Vertical Rock Faces in Pickens and Greenville Counties. *Castanea* 43(4):255–60.

McDermott, R. E., 1954. Seedling tolerance as a factor in bottomland timber succession. *Mo. Agr. Expt. Stat. Res. Bull.,* 557:1–11.

McVaugh, R., 1943. The vegetation of the Flat Granitic Outcrops of the Southeastern United States. *Ecol. Monogr.,* 13:121–66.

Melton, Frank A., 1950. The Carolina Bays. *Jour. of Geol.,* 58:128–34.

Mooring, M. T., A. W. Cooper and E. D. Seneca, 1971. Seed germination response and evidence for height ecophenes in *Spartina alterniflora* from North Carolina. *Am. Journal Bot.,* 48:143–72.

Morgan, Patrick H. 1974. *Study of tidelands and impoundments within a Three-river Delta System—The South Edisto, Ashepoo, and Combahee Rivers of South Carolina.* M.S. thesis, Univ. of Georgia, Athens.

Mowbray, Thomas B., 1966. Vegetational Gradients in the Bearwallow Gorge of the Blue Ridge Escarpment. *Jour. Elisha Mitchell Sci. Soc.* 82:138–49.

―――, and H. J. Oosting, 1968. Vegetation gradients in relation to environment and phenology in a southern Blue Ridge gorge. *Ecol. Monogr.,* 38:309–44.

Mullens, N. E., and C. L. Rodgers, 1964. Plants seldom seen

in South Carolina. *Castanea* 33(3):259– 60.

Nelson, T. C., 1955. Chestnut replacement in the southern highlands. *Ecology*, 36:352– 53.

———, 1957. The original forests of the Georgia Piedmont. *Ecology* 38:390– 97.

Nemeth, John C., 1968. The hardwood vegetation and soils of Hill Demonstration Forest, Durham County, North Carolina. *Jour. Elisha Mitchell Sci. Soc.*, 84:482– 91.

Odom, A. Leroy, and Paul D. Fullagar, 1973. Geochronologic and tectonic relationships between the inner piedmont, Brevard zone, and Blue Ridge belts, North Carolina. *Am. Jour. Sci.* 273-A:133– 49.

Oosting, H. J., 1942. An ecological analysis of the plant communities of Piedmont North Carolina. *Am. Midl. Nat.*, 28:1– 126.

———, 1945. Tolerance to salt spray of plants of coastal dunes. *Ecology* 26:85– 89.

———, 1954. Ecological processes and vegetation of the maritime strand in the Southeastern United States. *The Bot. Rev.*, 20:226.

———, and W. D. Billings, 1942. Factors effecting vegetational zonation on coastal dunes. *Ecology* 23:131– 42.

Overstreet, W. C., and Henry Bell, 1961. Geologic relations inferred from the provisional geologic map of the crystalline rocks of South Carolina. S.C. State Development Board, Division of Geology, *Geology Notes* 5: 39– 41.

——— and Henry Bell, 1965. The crystalline rocks of South Carolina. *U.S. Geology Survey Bulletin* 1183, p. 1– 126.

Peattie, R., 1943. *The Great Smokies and the Blue Ridge.* New York: Vanguard Press, 372 pp.

Penfound, W. T., 1952. Southern swamps and marshes. *Bot. Rev.*, 18:413– 46.

———, and T. D. Burleigh, 1941. Notes on the forest biology of Horn Island, Mississippi. *Ecology* 22:70– 78.

———, and J. R. Howard, 1940. A phytosociological study of an evergreen oak forest in the vicinity of New Orleans, Louisiana. *Amer. Midl. Nat.*, 23:165– 74.

———, and M. E. O'Neill, 1934. The vegetation of Cat Island, Mississippi. *Ecology* 15:1– 16.

Peterken, G. F., 1970. *Guide to the checksheet for I.B.P. areas.* Handbook No. 4. International Biological Program, London.

Phillips, J., 1931. Succession, development, the climax, and the complex organism: An analysis of concepts. *Jour. Ecol.*, 22:554– 71; 23:210– 46, 488– 508.

Pinson, J., 1973. *A floristic analysis of open dunes in South Carolina.* M.S. thesis, Univ. of South Carolina, Department of Biology.

Porcher, R. D., 1966. *A floristic study of the vascular plants in nine selected Carolina Bays in Berkeley County, South Carolina.* M.S. thesis Univ. of South Carolina.

Preston, F. W., 1948. The commonness, and rarity, of species. *Ecology*, 29:254– 83.

Prouty, W. F., 1950. Carolina Bays and their origin. *Jour. of Geol.*, 58:128– 34.

Purvis, John C., and Earl Rampey, 1975. *Weather Extremes in South Carolina.* S.C. Disaster Preparedness Agency.

Quarterman, E., and C. Keever, 1962. Southern mixed hardwood forest: climax in the southeastern coastal plain. *Ecol. Mongr.*, 32:167– 85.

Racine, C. H., 1966. Pine communities and their site characteristics in the Blue Ridge escarpment. *Jour. Elisha Mitchell Sci. Soc.*, 82:172– 81.

Radford, A. E., 1959. A relic plant community in South Carolina. *Jour. Elisha Mitchell Sci. Soc.*, 75:33– 34.

———, 1974. *Field data and information on plant communities in the Eastern United States.* Unpublished.

Rayner, D. A., 1976. *A Monograph concerning the water elm Planera Aquatica (Walt.) J. F. Gmelin (Ulmaceae).* Ph.D. dissertation, Dept. of Biology, Univ. of South Carolina.

———, 1978. Rare blueberry to be nominated for national endangered status. *The Heritage Trust Newsletter* 2(3). S.C. Wildlife and Marine Resources Department.

————, and W. T. Batson, 1976. Maritime closed dunes vegetation in South Carolina. *Castanea* 41(1):58–70.

Reed, John F., 1939. Root and shoot growth of shortleaf and loblolly pines in relation to certain environmental conditions. *Duke Univ. School of Forestry Bull.*, 4:52.

Robinson, P., 1954. The distribution of plant populations. *Ann. Bot.*, 18:35–46.

Rodgers, C. Leland, 1965. Vegetation of Horsepasture Gorge. *Jour. Elisha Mitchell Sci. Soc.*, 81:103–12.

————, 1955. Vascular plants of Table Rock Mountain, South Carolina. *Castanea* 20:122–44.

————, 1969. Vascular plants in Horsepasture Gorge. *Castanea* 34:374.

————, and N. E. Mullens, 1969. State records for upper South Carolina. *Castanea* 38:114–16.

————, and Roy E. Shake, 1965. A survey of vascular plants in Bearcamp Creek Watershed. *Castanea* 30(3):149–66.

Rodgers, John, 1972. Latest Precambrian (post Grenville) rocks of the Appalachian Region. *Am. Jour. Sci.* 272:507–20.

Roper, J. Paul, and David E. Dunn, 1970. *Geology of the Tamassee, Satolah and Cashiers quadrangles, Oconee County, South Carolina.* S.C. State Development Board, Division of Geology.

Rothrock, Paul E., and Richard H. Wagner, 1975. *Eleocharis acicularis* (L.) R. & S.: The Autecology of an acid tolerant sedge. *Castanea* 40(4):279.

Sargent, C. S., 1886. Some remarks upon the journey of André Michaux to the high mountains of Carolina, in December 1788, in a letter addressed to Professor Asa Gray. *Am. Jour. Sci.*, ser. 3, 22:466–73.

Seaborn, Margaret Mills, Ed., 1976. *André Michaux's journeys in Oconee County.* R. L. Bryan Company, Columbia, S.C.

Secor, D. T., and Wagener, H. D., 1968. Stratigraphy, structure and petrology of the piedmont in central South Carolina. Carolina Geological Society field trip; in S.C. State Development Board, Division of Geology, *Geologic Notes* 12: 67–84.

Shelford, V. E., 1963. *The Ecology of North America.* Univ. of Illinois Press.

Shirley, H. L., 1935. Light as an ecological factor and its measurement. *Bot. Rev.*, 1:355–81.

Shreve, F., 1917. A map of the vegetation of the United States. *Geog. Rev.* 3:119–25.

Siple, George E., 1960. Some geologic and hydrologic factors affecting limestone terraces of Tertiary age in South Carolina. *Southeastern Geol.*, 2(1):1–11.

————, 1975. *Ground-water Resources of Orangeburg County, South Carolina.* U.S. Geological Survey.

————, 1967. *Geology and ground water on the Savannah River Plant and Vicinity.* U.S. Geol. Survey Water-Supply Paper 1841, 113 pp.

Sloan, Earle, 1908. *Catalogue of the mineral localities of South Carolina.* S.C. State Development Board, Division of Geology, Bull., 2, 505, pp.

Smith, L. L., 1961. Historical map links progress in geology. S.C. State Development Board, Division of Geology, *Geology Notes*, v. 5, pp. 93–96.

Smith, R. L., 1966. *Ecology and field biology.* Harper and Row.

Society of American Foresters, 1952. *Forest cover types of North America.* Society of American Foresters, Washington, D.C.

Stalter, Richard, 1973a. Transplantation of Salt marsh vegetation II., Georgetown, South Carolina. *Castanea* 38:132–39.

————, 1973b. Factors influencing the distribution of vegetation of the Cooper River estuary. *Castanea* 38(1):18–24.

————, 1974a. Vegetation in coastal dunes of South Carolina. *Castanea* 39(1):95–103.

————, 1974b. *The vegetation of the Cooper River estuary*, in Cooper River Environmental Study, S.C. Water Resources Commission.

————, 1975. The flora of the Isle of Palms, S.C. *Castanea* 40:4–13.

————, and W. T. Batson, 1973. Seed viability in salt marsh

taxa, Georgetown County, South Carolina. *Castanea* 38(1):109–13.

Swanton, J. R., 1946. The Indians of the southeastern United States. *Bur. Am. Ethnol. Bull.* 137, 943 pp.

Tansley, A. G., 1920. The classification of vegetation and the concept of development. *Jour. Ecol.*, 8:118–49.

Thornthwaite, C. W., 1931. The climates of North America. *Geog. Rev.*, 21:633–54.

U.S. Dept. of Agriculture Soil Conservation Service, 1962. *Soils Memorandum SC-4.*

Wagener, H. D., 1970. *Geology of the southern two-thirds of the Winnsboro 15-minute quadrangle, South Carolina.* S.C. State Development Board, Division of Geology.

Wagner, R. H., 1964. The ecology of *Uniola paniculata* L., in the dune strand habitat of North Carolina. *Ecological Monogr.* 34:79–96.

Welch, Richard L., 1968. *Forest statistics for the northern coastal plain of South Carolina.* Southeast. Forest Exp. Sta., U.S. Forest Serv. Resource Bull. SE-10, 35 pp.

Wells, B. W., 1924. *Major Plant communities of North Carolina.* Tech. Bull. of the N.C. Agric. Exp. Station, Tech. Bull. No.25.

———, 1928. Plant communities of the coastal plain of North Carolina and their successional relations. *Ecology*, 9:230–42.

———, 1939. A new forest climax; the salt spray climax of Smith Island, North Carolina. *Bull. Torr. Bot. Club*, 66:629–34.

———, 1967. *The natural gardens of North Carolina.* Univ. of North Carolina Press, Chapel Hill.

———, and S. G. Boyce, 1953. Carolina bays: Additional data on their origin, age, and history. *Jour. Elisha Mitchell Sci. Soc.*, 69:119–40.

———, and I. V. Shunk, 1931. The vegetation and habitat factors of the coarse sands of the North Carolina coastal plain: An ecological study. *Ecol. Monogr.* 1:465–520.

———, and I. V. Shunk, 1937. Seaside Shrubs: wind forms vs. spray forms. *Science* 85:499.

———, and I. V. Shunk, 1938. Salt spray: an important factor in coastal ecology. *Bull. Torr. Bot. Club*, 65:485–92.

Whittaker, R. H., 1951. A criticism of the plant association and climatic climax concepts. *Northwest Sci.*, 25:17–31.

Woodwell, George M., 1956. Vegetation of the Great

Smokey Mountains. *Ecol. Monogr.* 26:1–80.

———, 1958. Factors controlling growth of pond pine seedlings in organic soil of the Carolinas. *Ecol. Monogr.*, 28:219–35.

Wright, A. H., and A. A. Wright, 1932. The habitats and composition of the vegetation of Okefenokee Swamp, Georgia. *Ecol. Monogr.*, 2:190–323.

General Index and
Index to Scientific Plant Names

Index to Common Plant Names